The Standard™

PILOT MASTER LOG

AVIATION SUPPLIES & ACADEMICS
NEWCASTLE, WASHINGTON

The Standard™ Pilot Master Log
SP-6

ASA-SP-6
ISBN 978-1-56027-889-4

Published by
Aviation Supplies & Academics, Inc.
7005 132nd Place SE
Newcastle, Washington 98059 USA
Website: asa2fly.com
Email: asa@asa2fly.com

Printed in China

[18] 21

TRANSPORTATION USD $26.95

ISBN 978-1-56027-889-4

52695 >

9 781560 278894

Name ______________________________

Address ______________________________

____

Telephone ______________________________

Book No. __________

From Date __________

Hours __________

To Date __________

Hours __________

LICENSE				RATING			
Date of Issue	Type of License/Number	Examiner/Authority	Remarks	Date of Issue	Type of License/Number	Examiner/Authority	Remarks

Change of Address

Effective as of

Contact Information

Effective as of

Home

Business

Fax

Cell

Email

Home

Business

Fax

Cell

Email

Home

Business

Fax

Cell

Email

Instructions for Use

1. Logbook entries are to be made as soon as possible upon the completion of a flight. Every entry and signature are made in ink or indelible pencil. The logbook holder should keep current the address, contact information, medical history and employer information during the currency of the logbook.
2. The logbook holder should have the most recent logbook readily available for inspection upon request by an authorized representative of the controlling authority. The holder might also need to produce the logbook when obtaining a new license or rating and for renewal and revalidation of existing licenses and ratings.
3. Columns labeled Examiner/Authority or Examiner/Ck. Pilot/Authority may be used for the name or number, as applicable, of the official providing the endorsement.
4. The specifics of every flight during which the holder acts as a member of a flight crew are to be recorded in this logbook, using one line per flight. However, if an aircraft carries out a series of flights on the same day, returning on each occasion to the same place of departure and the interval between successive flights does not exceed thirty minutes, such a series of flights may be recorded as a single entry and noted in the Remarks column.
5. Each flight should show departure and arrival points. A flight not landing away from the departure point may be written in the To column as “local” or “LCL.”
6. Record the place of departure and arrival using the place name in full or the internationally recognized three- or four-letter designator.
7. Departure time is the time the aircraft first moves under its own power. Arrival time is the time when the aircraft comes to rest after landing. Helicopter flight time is from the first movement under its own power until the rotors are stopped.
8. Times are to be recorded in the Coordinated Universal Time (UTC). Flying time may be recorded as hours and minutes (HH:MM) or as hours and decimals (HH.DD) to the nearest two decimals of an hour (see the Time Conversion Table in the back of the logbook).
9. When an aircraft carries more than one pilot as a member of the required flight crew, one member shall be designated as the Pilot In Command (Captain or Commander) of the flight and his/her name shall be entered in the “Remarks, Procedures, Maneuvers, Endorsements, PIC Name” column. The logbook holder enters “Self” when s/he is the Pilot In Command.
10. If the flight occurred in an aircraft other than a Single Engine (SE) or Multi Engine (ME), enter the appropriate aircraft category or class under the appropriate heading before entering the flight time.
11. The total session time in the Synthetic Flight Training Device (FTD) should include all exercises carried out in the trainer. Enter the type and qualification number of the device in the appropriate column.
12. The following system of notation may be used for entries in the columns headed “Pilot Function”:

Operating Capacity	Designation	Flight Time recorded in column:
Pilot In Command	PIC or P1	PIC
Student Pilot In Command	SPIC	PIC
Pilot In Command Under Supervision	PICUS	PIC
Co-Pilot	P2	Co-Pilot
Pilot Under Instruction	P/UT	Dual Received
Flight Instr./Flight Ex./Ck. Pilot	FI/FE/CK	Instructor/Examiner/Ck. Pilot

13. The “Remarks” column is used to record brief comments and references to endorsements, weather, routings, flight training exercises, specific PIC designation (such as SPIC, PICUS, FI, FE, Ck. Pilot), synthetic flight training device instruction and exercises, consecutive series of flights not greater than 30 minutes, etc.
14. At the completion of each page, the columns should be totaled, including any figures “brought forward” as appropriate and certified by the logbook holder. The new totals are carried forward to the next page.
15. Pages for compiling annual or periodic summaries of flight time shall be used to record all time relevant and/or required to maintain currency of experience for specific types of aircraft or flying conditions, such as those used to satisfy professionally mandated standards.
16. At the bottom of the summary pages, there is space provided for physical examinations and flight checks so that relevant information for a flight review may be conveniently reviewed.

Proficiency and Type Ratings

FLIGHT PROFICIENCY, CHECKS AND REVIEWS				TYPE RATING					
Date	Type of Proficiency, Check or Review	Examiner/Ck. Pilot/Authority	Remarks	Date of Issue	Type of Aircraft	Number	Engine(s) Make and Model	Examiner/Authority	Remarks

Medical History and Employer Information

MEDICAL HISTORY					EMPLOYER INFORMATION		
Date	Type of Medical	Number	Examiner/Authority	Remarks	Name and Address of Employer	From	To

Year 20__ Date	Aircraft Type (Make, Model, Variant)	Aircraft Ident. Mark, Reg. No.	Departure & Arrival		Miles Flown	Flight No.	Number of		No. Instr. Appr.	Aircraft Category and Class				Pilot Function			
			From Time	To Time			TO Day Night	LDG Day Night		SE	ME			PIC (Solo, SPIC, PICUS)	Co-Pilot	Dual Received	Instructor/ Examiner/ Ck. Pilot
						Page Total											
						Amount Forward											
						Total to Date											

	Operational Conditions					Synthetic Flight				
	Cross-Country	Night	Actual Instrument	Instrument (Hood)		Time of Session	Type of Instruction	Type and Device No.	Total Duration of Flight	Remarks, Procedures, Maneuvers, Endorsements, PIC Name

I certify that the statements made by me on this form are true.

Pilot's Signature ____________________

Year 20__ Date	Aircraft Type (Make, Model, Variant)	Aircraft Ident. Mark, Reg. No.	Departure & Arrival From Time	To Time	Miles Flown	Flight No.	Number of TO Day Night	LDG Day Night	No. Instr. Appr.	Aircraft Category and Class SE	ME			Pilot Function PIC (Solo, SPIC, PICUS)	Co-Pilot	Dual Received	Instructor/ Examiner/ Ck. Pilot
						Page Total											
						Amount Forward											
						Total to Date											

		Operational Conditions										Synthetic Flight					
		Cross-Country		Night		Actual Instrument		Instrument (Hood)				Time of Session	Type of Instruction	Type and Device No.	Total Duration of Flight		Remarks, Procedures, Maneuvers, Endorsements, PIC Name

I certify that the statements made by me on this form are true.

Pilot's Signature ______________________________

Year			Departure & Arrival				Number of			Aircraft Category and Class								Pilot Function							
20__ Date	Aircraft Type (Make, Model, Variant)	Aircraft Ident. Mark, Reg. No.	From Time	To Time	Miles Flown	Flight No.	TO Day Night	LDG Day Night	No. Instr. Appr.	SE		ME						PIC (Solo, SPIC, PICUS)		Co-Pilot		Dual Received		Instructor/ Examiner/ Ck. Pilot	
						Page Total																			
						Amount Forward																			
						Total to Date																			

		Operational Conditions										Synthetic Flight					
		Cross-Country		Night		Actual Instrument		Instrument (Hood)				Time of Session	Type of Instruction	Type and Device No.	Total Duration of Flight		Remarks, Procedures, Maneuvers, Endorsements, PIC Name

I certify that the statements made by me on this form are true.

Pilot's Signature ______________________

Year 20__ Date	Aircraft Type (Make, Model, Variant)	Aircraft Ident. Mark, Reg. No.	Departure & Arrival: From	Time	To	Time	Miles Flown	Flight No.	Number of: TO Day	Night	LDG Day	Night	No. Instr. Appr.	Aircraft Category and Class: SE		ME						Pilot Function: PIC (Solo, SPIC, PICUS)		Co-Pilot		Dual Received		Instructor/ Examiner/ Ck. Pilot	
								Page Total																					
								Amount Forward																					
								Total to Date																					

	Operational Conditions					Synthetic Flight				
	Cross-Country	Night	Actual Instrument	Instrument (Hood)		Time of Session	Type of Instruction	Type and Device No.	Total Duration of Flight	Remarks, Procedures, Maneuvers, Endorsements, PIC Name

I certify that the statements made by me on this form are true.

Pilot's Signature ______________________

Year 20__ Date	Aircraft Type (Make, Model, Variant)	Aircraft Ident. Mark, Reg. No.	Departure & Arrival: From / Time	Departure & Arrival: To / Time	Miles Flown	Flight No.	Number of: TO Day / Night	Number of: LDG Day / Night	No. Instr. Appr.	Aircraft Category and Class: SE	ME			Pilot Function: PIC (Solo, SPIC, PICUS)	Co-Pilot	Dual Received	Instructor/ Examiner/ Ck. Pilot
					Page Total												
					Amount Forward												
					Total to Date												

		Operational Conditions										Synthetic Flight					
		Cross-Country		Night		Actual Instrument		Instrument (Hood)				Time of Session	Type of Instruction	Type and Device No.	Total Duration of Flight		Remarks, Procedures, Maneuvers, Endorsements, PIC Name

I certify that the statements made by me on this form are true.

Pilot's Signature ______________________________

Year 20__ Date	Aircraft Type (Make, Model, Variant)	Aircraft Ident. Mark, Reg. No.	**Departure & Arrival**		Miles Flown	Flight No.	Number of		No. Instr. Appr.	**Aircraft Category and Class**								**Pilot Function**							
			From Time	To Time			TO Day Night	LDG Day Night		SE		ME						PIC (Solo, SPIC, PICUS)		Co-Pilot		Dual Received		Instructor/ Examiner/ Ck. Pilot	
						Page Total																			
						Amount Forward																			
						Total to Date																			

	Operational Conditions					Synthetic Flight				
	Cross-Country	Night	Actual Instrument	Instrument (Hood)		Time of Session	Type of Instruction	Type and Device No.	Total Duration of Flight	Remarks, Procedures, Maneuvers, Endorsements, PIC Name

I certify that the statements made by me on this form are true.

Pilot's Signature ______________________

Year 20__ Date	Aircraft Type (Make, Model, Variant)	Aircraft Ident. Mark, Reg. No.	Departure & Arrival		Miles Flown	Flight No.	Number of		No. Instr. Appr.	Aircraft Category and Class				Pilot Function			
			From Time	To Time			TO Day Night	LDG Day Night		SE	ME			PIC (Solo, SPIC, PICUS)	Co-Pilot	Dual Received	Instructor/ Examiner/ Ck. Pilot
						Page Total											
						Amount Forward											
						Total to Date											

	Operational Conditions					Synthetic Flight				
	Cross-Country	Night	Actual Instrument	Instrument (Hood)		Time of Session	Type of Instruction	Type and Device No.	Total Duration of Flight	Remarks, Procedures, Maneuvers, Endorsements, PIC Name

I certify that the statements made by me on this form are true.

Pilot's Signature ______________________________

Year 20__ Date	Aircraft Type (Make, Model, Variant)	Aircraft Ident. Mark, Reg. No.	**Departure & Arrival**		Miles Flown	Flight No.	Number of		No. Instr. Appr.	**Aircraft Category and Class**								**Pilot Function**							
			From Time	To Time			TO Day Night	LDG Day Night		SE		ME						PIC (Solo, SPIC, PICUS)		Co-Pilot		Dual Received		Instructor/ Examiner/ Ck. Pilot	
						Page Total																			
						Amount Forward																			
						Total to Date																			

	Operational Conditions					Synthetic Flight				
	Cross-Country	Night	Actual Instrument	Instrument (Hood)		Time of Session	Type of Instruction	Type and Device No.	Total Duration of Flight	Remarks, Procedures, Maneuvers, Endorsements, PIC Name

I certify that the statements made by me on this form are true.

Pilot's Signature ____________________

Year 20__ Date	Aircraft Type (Make, Model, Variant)	Aircraft Ident. Mark, Reg. No.	Departure & Arrival From Time	To Time	Miles Flown	Flight No.	Number of TO Day Night	LDG Day Night	No. Instr. Appr.	Aircraft Category and Class SE	ME			Pilot Function PIC (Solo, SPIC, PICUS)	Co-Pilot	Dual Received	Instructor/ Examiner/ Ck. Pilot
							Page Total										
							Amount Forward										
							Total to Date										

	Operational Conditions					Synthetic Flight				
	Cross-Country	Night	Actual Instrument	Instrument (Hood)		Time of Session	Type of Instruction	Type and Device No.	Total Duration of Flight	Remarks, Procedures, Maneuvers, Endorsements, PIC Name

I certify that the statements made by me on this form are true.

Pilot's Signature ______________________________

Year 20__ Date	Aircraft Type (Make, Model, Variant)	Aircraft Ident. Mark, Reg. No.	Departure & Arrival		Miles Flown	Flight No.	Number of		No. Instr. Appr.	Aircraft Category and Class				Pilot Function			
			From Time	To Time			TO Day Night	LDG Day Night		SE	ME			PIC (Solo, SPIC, PICUS)	Co-Pilot	Dual Received	Instructor/ Examiner/ Ck. Pilot
						Page Total											
						Amount Forward											
						Total to Date											

		Operational Conditions										Synthetic Flight					
		Cross-Country		Night		Actual Instrument		Instrument (Hood)				Time of Session	Type of Instruction	Type and Device No.	Total Duration of Flight		Remarks, Procedures, Maneuvers, Endorsements, PIC Name

I certify that the statements made by me on this form are true.

Pilot's Signature ______________________________

Year 20__ Date	Aircraft Type (Make, Model, Variant)	Aircraft Ident. Mark, Reg. No.	Departure & Arrival		Miles Flown	Flight No.	Number of		No. Instr. Appr.	Aircraft Category and Class				Pilot Function			
			From Time	To Time			TO Day / Night	LDG Day / Night		SE	ME			PIC (Solo, SPIC, PICUS)	Co-Pilot	Dual Received	Instructor/ Examiner/ Ck. Pilot
						Page Total											
						Amount Forward											
						Total to Date											

	Operational Conditions					Synthetic Flight				
	Cross-Country	Night	Actual Instrument	Instrument (Hood)		Time of Session	Type of Instruction	Type and Device No.	Total Duration of Flight	Remarks, Procedures, Maneuvers, Endorsements, PIC Name

I certify that the statements made by me on this form are true.

Pilot's Signature ______________________

Year 20__ Date	Aircraft Type (Make, Model, Variant)	Aircraft Ident. Mark, Reg. No.	Departure & Arrival		Miles Flown	Flight No.	Number of		No. Instr. Appr.	Aircraft Category and Class								Pilot Function							
			From Time	To Time			TO Day Night	LDG Day Night		SE		ME						PIC (Solo, SPIC, PICUS)		Co-Pilot		Dual Received		Instructor/ Examiner/ Ck. Pilot	
Page Total																									
Amount Forward																									
Total to Date																									

	Operational Conditions					Synthetic Flight				
	Cross-Country	Night	Actual Instrument	Instrument (Hood)		Time of Session	Type of Instruction	Type and Device No.	Total Duration of Flight	Remarks, Procedures, Maneuvers, Endorsements, PIC Name

I certify that the statements made by me on this form are true.

Pilot's Signature ______________________________

Year 20__ Date	Aircraft Type (Make, Model, Variant)	Aircraft Ident. Mark, Reg. No.	**Departure & Arrival** From Time	To Time	Miles Flown	Flight No.	Number of TO Day Night	LDG Day Night	No. Instr. Appr.	**Aircraft Category and Class** SE	ME			**Pilot Function** PIC (Solo, SPIC, PICUS)	Co-Pilot	Dual Received	Instructor/ Examiner/ Ck. Pilot
						Page Total											
						Amount Forward											
						Total to Date											

		Operational Conditions										Synthetic Flight					
		Cross-Country		Night		Actual Instrument		Instrument (Hood)				Time of Session	Type of Instruction	Type and Device No.	Total Duration of Flight		Remarks, Procedures, Maneuvers, Endorsements, PIC Name

I certify that the statements made by me on this form are true.

Pilot's Signature ______________________________

Year			Departure & Arrival				Number of		No.	Aircraft Category and Class								Pilot Function							
20__ Date	Aircraft Type (Make, Model, Variant)	Aircraft Ident. Mark, Reg. No.	From Time	To Time	Miles Flown	Flight No.	TO Day Night	LDG Day Night	Instr. Appr.	SE		ME						PIC (Solo, SPIC, PICUS)		Co-Pilot		Dual Received		Instructor/ Examiner/ Ck. Pilot	
							Page Total																		
							Amount Forward																		
							Total to Date																		

		Operational Conditions										Synthetic Flight					
		Cross-Country		Night		Actual Instrument		Instrument (Hood)				Time of Session	Type of Instruction	Type and Device No.	Total Duration of Flight		Remarks, Procedures, Maneuvers, Endorsements, PIC Name

I certify that the statements made by me on this form are true.

Pilot's Signature ______________________________

Year 20__ Date	Aircraft Type (Make, Model, Variant)	Aircraft Ident. Mark, Reg. No.	**Departure & Arrival**		Miles Flown	Flight No.	Number of		No. Instr. Appr.	**Aircraft Category and Class**				**Pilot Function**			
			From Time	To Time			TO Day Night	LDG Day Night		SE	ME			PIC (Solo, SPIC, PICUS)	Co-Pilot	Dual Received	Instructor/ Examiner/ Ck. Pilot
						Page Total											
						Amount Forward											
						Total to Date											

		Operational Conditions										Synthetic Flight					
		Cross-Country		Night		Actual Instrument		Instrument (Hood)				Time of Session	Type of Instruction	Type and Device No.	Total Duration of Flight		Remarks, Procedures, Maneuvers, Endorsements, PIC Name

I certify that the statements made by me on this form are true.

Pilot's Signature ______________________________

Year 20__ Date	Aircraft Type (Make, Model, Variant)	Aircraft Ident. Mark, Reg. No.	Departure & Arrival		Miles Flown	Flight No.	Number of		No. Instr. Appr.	Aircraft Category and Class								Pilot Function							
			From Time	To Time			TO Day Night	LDG Day Night		SE		ME						PIC (Solo, SPIC, PICUS)		Co-Pilot		Dual Received		Instructor/ Examiner/ Ck. Pilot	
						Page Total																			
						Amount Forward																			
						Total to Date																			

		Operational Conditions										Synthetic Flight					
		Cross-Country		Night		Actual Instrument		Instrument (Hood)				Time of Session	Type of Instruction	Type and Device No.	Total Duration of Flight		Remarks, Procedures, Maneuvers, Endorsements, PIC Name

I certify that the statements made by me on this form are true.

Pilot's Signature ______________________

Year 20__ Date	Aircraft Type (Make, Model, Variant)	Aircraft Ident. Mark, Reg. No.	**Departure & Arrival**		Miles Flown	Flight No.	Number of		No. Instr. Appr.	**Aircraft Category and Class**				**Pilot Function**			
			From Time	To Time			TO Day Night	LDG Day Night		SE	ME			PIC (Solo, SPIC, PICUS)	Co-Pilot	Dual Received	Instructor/ Examiner/ Ck. Pilot
							Page Total										
							Amount Forward										
							Total to Date										

	Operational Conditions					Synthetic Flight				
	Cross-Country	Night	Actual Instrument	Instrument (Hood)		Time of Session	Type of Instruction	Type and Device No.	Total Duration of Flight	Remarks, Procedures, Maneuvers, Endorsements, PIC Name

I certify that the statements made by me on this form are true.

Pilot's Signature ______________________________

Year 20__ Date	Aircraft Type (Make, Model, Variant)	Aircraft Ident. Mark, Reg. No.	Departure & Arrival		Miles Flown	Flight No.	Number of		No. Instr. Appr.	Aircraft Category and Class				Pilot Function			
			From Time	To Time			TO Day Night	LDG Day Night		SE	ME			PIC (Solo, SPIC, PICUS)	Co-Pilot	Dual Received	Instructor/ Examiner/ Ck. Pilot
						Page Total											
						Amount Forward											
						Total to Date											

	Operational Conditions					Synthetic Flight				
	Cross-Country	Night	Actual Instrument	Instrument (Hood)		Time of Session	Type of Instruction	Type and Device No.	Total Duration of Flight	Remarks, Procedures, Maneuvers, Endorsements, PIC Name

I certify that the statements made by me on this form are true.

Pilot's Signature ______________________

Year 20__ Date	Aircraft Type (Make, Model, Variant)	Aircraft Ident. Mark, Reg. No.	**Departure & Arrival**		Miles Flown	Flight No.	Number of		No. Instr. Appr.	**Aircraft Category and Class**				**Pilot Function**			
			From Time	To Time			TO Day Night	LDG Day Night		SE	ME			PIC (Solo, SPIC, PICUS)	Co-Pilot	Dual Received	Instructor/ Examiner/ Ck. Pilot
						Page Total											
						Amount Forward											
						Total to Date											

	Operational Conditions					Synthetic Flight				
	Cross-Country	Night	Actual Instrument	Instrument (Hood)		Time of Session	Type of Instruction	Type and Device No.	Total Duration of Flight	Remarks, Procedures, Maneuvers, Endorsements, PIC Name

I certify that the statements made by me on this form are true.

Pilot's Signature ______________________________

Year 20__ Date	Aircraft Type (Make, Model, Variant)	Aircraft Ident. Mark, Reg. No.	Departure & Arrival		Miles Flown	Flight No.	Number of		No. Instr. Appr.	Aircraft Category and Class								Pilot Function							
			From Time	To Time			TO Day Night	LDG Day Night		SE		ME						PIC (Solo, SPIC, PICUS)		Co-Pilot		Dual Received		Instructor/ Examiner/ Ck. Pilot	
						Page Total																			
						Amount Forward																			
						Total to Date																			

	Operational Conditions					Synthetic Flight				
	Cross-Country	Night	Actual Instrument	Instrument (Hood)		Time of Session	Type of Instruction	Type and Device No.	Total Duration of Flight	Remarks, Procedures, Maneuvers, Endorsements, PIC Name

I certify that the statements made by me on this form are true.

Pilot's Signature ______________________________

Year 20__			Departure & Arrival				Number of			Aircraft Category and Class								Pilot Function							
Date	Aircraft Type (Make, Model, Variant)	Aircraft Ident. Mark, Reg. No.	From Time	To Time	Miles Flown	Flight No.	TO Day Night	LDG Day Night	No. Instr. Appr.	SE		ME						PIC (Solo, SPIC, PICUS)		Co-Pilot		Dual Received		Instructor/ Examiner/ Ck. Pilot	
							Page Total																		
							Amount Forward																		
							Total to Date																		

		Operational Conditions										Synthetic Flight					
		Cross-Country		Night		Actual Instrument		Instrument (Hood)				Time of Session	Type of Instruction	Type and Device No.	Total Duration of Flight		Remarks, Procedures, Maneuvers, Endorsements, PIC Name

I certify that the statements made by me on this form are true.

Pilot's Signature ______________________________

Year 20__ Date	Aircraft Type (Make, Model, Variant)	Aircraft Ident. Mark, Reg. No.	Departure & Arrival		Miles Flown	Flight No.	Number of		No. Instr. Appr.	Aircraft Category and Class				Pilot Function			
			From Time	To Time			TO Day Night	LDG Day Night		SE	ME			PIC (Solo, SPIC, PICUS)	Co-Pilot	Dual Received	Instructor/ Examiner/ Ck. Pilot
						Page Total											
						Amount Forward											
						Total to Date											

	Operational Conditions					Synthetic Flight				
	Cross-Country	Night	Actual Instrument	Instrument (Hood)		Time of Session	Type of Instruction	Type and Device No.	Total Duration of Flight	Remarks, Procedures, Maneuvers, Endorsements, PIC Name

I certify that the statements made by me on this form are true.

Pilot's Signature ______________________________

Year 20__ Date	Aircraft Type (Make, Model, Variant)	Aircraft Ident. Mark, Reg. No.	Departure & Arrival		Miles Flown	Flight No.	Number of		No. Instr. Appr.	Aircraft Category and Class								Pilot Function							
			From Time	To Time			TO Day Night	LDG Day Night		SE		ME						PIC (Solo, SPIC, PICUS)		Co-Pilot		Dual Received		Instructor/ Examiner/ Ck. Pilot	
						Page Total																			
						Amount Forward																			
						Total to Date																			

	Operational Conditions					Synthetic Flight				
	Cross-Country	Night	Actual Instrument	Instrument (Hood)		Time of Session	Type of Instruction	Type and Device No.	Total Duration of Flight	Remarks, Procedures, Maneuvers, Endorsements, PIC Name

I certify that the statements made by me on this form are true.

Pilot's Signature ______________________

Year 20__ Date	Aircraft Type (Make, Model, Variant)	Aircraft Ident. Mark, Reg. No.	**Departure & Arrival**		Miles Flown	Flight No.	Number of		No. Instr. Appr.	**Aircraft Category and Class**								**Pilot Function**							
			From Time	To Time			TO Day Night	LDG Day Night		SE		ME						PIC (Solo, SPIC, PICUS)		Co-Pilot		Dual Received		Instructor/ Examiner/ Ck. Pilot	
						Page Total																			
						Amount Forward																			
						Total to Date																			

	Operational Conditions					Synthetic Flight				
	Cross-Country	Night	Actual Instrument	Instrument (Hood)		Time of Session	Type of Instruction	Type and Device No.	Total Duration of Flight	Remarks, Procedures, Maneuvers, Endorsements, PIC Name

I certify that the statements made by me on this form are true.

Pilot's Signature ______________________

Year 20__ Date	Aircraft Type (Make, Model, Variant)	Aircraft Ident. Mark, Reg. No.	Departure & Arrival		Miles Flown	Flight No.	Number of		No. Instr. Appr.	Aircraft Category and Class				Pilot Function			
			From Time	To Time			TO Day Night	LDG Day Night		SE	ME			PIC (Solo, SPIC, PICUS)	Co-Pilot	Dual Received	Instructor/ Examiner/ Ck. Pilot
						Page Total											
						Amount Forward											
						Total to Date											

		Operational Conditions										Synthetic Flight					
		Cross-Country		Night		Actual Instrument		Instrument (Hood)				Time of Session	Type of Instruction	Type and Device No.	Total Duration of Flight		Remarks, Procedures, Maneuvers, Endorsements, PIC Name

I certify that the statements made by me on this form are true.

Pilot's Signature ______________________________

Year 20__ Date	Aircraft Type (Make, Model, Variant)	Aircraft Ident. Mark, Reg. No.	**Departure & Arrival**		Miles Flown	Flight No.	Number of		No. Instr. Appr.	**Aircraft Category and Class**				**Pilot Function**			
			From Time	To Time			TO Day Night	LDG Day Night		SE	ME			PIC (Solo, SPIC, PICUS)	Co-Pilot	Dual Received	Instructor/ Examiner/ Ck. Pilot
						Page Total											
						Amount Forward											
						Total to Date											

	Operational Conditions					Synthetic Flight				
	Cross-Country	Night	Actual Instrument	Instrument (Hood)		Time of Session	Type of Instruction	Type and Device No.	Total Duration of Flight	Remarks, Procedures, Maneuvers, Endorsements, PIC Name

I certify that the statements made by me on this form are true.

Pilot's Signature ______________________________

Year 20__ Date	Aircraft Type (Make, Model, Variant)	Aircraft Ident. Mark, Reg. No.	Departure & Arrival		Miles Flown	Flight No.	Number of		No. Instr. Appr.	Aircraft Category and Class				Pilot Function			
			From Time	To Time			TO Day Night	LDG Day Night		SE	ME			PIC (Solo, SPIC, PICUS)	Co-Pilot	Dual Received	Instructor/ Examiner/ Ck. Pilot
						Page Total											
						Amount Forward											
						Total to Date											

	Operational Conditions					Synthetic Flight				
	Cross-Country	Night	Actual Instrument	Instrument (Hood)		Time of Session	Type of Instruction	Type and Device No.	Total Duration of Flight	Remarks, Procedures, Maneuvers, Endorsements, PIC Name

I certify that the statements made by me on this form are true.

Pilot's Signature ____________________

Year 20__ Date	Aircraft Type (Make, Model, Variant)	Aircraft Ident. Mark, Reg. No.	Departure & Arrival From Time	To Time	Miles Flown	Flight No.	Number of TO Day / Night	LDG Day / Night	No. Instr. Appr.	Aircraft Category and Class SE		ME						Pilot Function PIC (Solo, SPIC, PICUS)		Co-Pilot		Dual Received		Instructor/ Examiner/ Ck. Pilot	
							Page Total																		
							Amount Forward																		
							Total to Date																		

	Operational Conditions					Synthetic Flight				
	Cross-Country	Night	Actual Instrument	Instrument (Hood)		Time of Session	Type of Instruction	Type and Device No.	Total Duration of Flight	Remarks, Procedures, Maneuvers, Endorsements, PIC Name

I certify that the statements made by me on this form are true.

Pilot's Signature ______________________________

Year 20__ Date	Aircraft Type (Make, Model, Variant)	Aircraft Ident. Mark, Reg. No.	**Departure & Arrival**		Miles Flown	Flight No.	Number of		No. Instr. Appr.	**Aircraft Category and Class**								**Pilot Function**							
			From Time	To Time			TO Day Night	LDG Day Night		SE		ME						PIC (Solo, SPIC, PICUS)		Co-Pilot		Dual Received		Instructor/ Examiner/ Ck. Pilot	
						Page Total																			
						Amount Forward																			
						Total to Date																			

	Operational Conditions					Synthetic Flight				
	Cross-Country	Night	Actual Instrument	Instrument (Hood)		Time of Session	Type of Instruction	Type and Device No.	Total Duration of Flight	Remarks, Procedures, Maneuvers, Endorsements, PIC Name

I certify that the statements made by me on this form are true.

Pilot's Signature ____________________

Year 20__ Date	Aircraft Type (Make, Model, Variant)	Aircraft Ident. Mark, Reg. No.	Departure & Arrival		Miles Flown	Flight No.	Number of		No. Instr. Appr.	Aircraft Category and Class				Pilot Function			
			From Time	To Time			TO Day Night	LDG Day Night		SE	ME			PIC (Solo, SPIC, PICUS)	Co-Pilot	Dual Received	Instructor/ Examiner/ Ck. Pilot
						Page Total											
						Amount Forward											
						Total to Date											

		Operational Conditions										Synthetic Flight					
		Cross-Country		Night		Actual Instrument		Instrument (Hood)				Time of Session	Type of Instruction	Type and Device No.	Total Duration of Flight		Remarks, Procedures, Maneuvers, Endorsements, PIC Name

I certify that the statements made by me on this form are true.

Pilot's Signature ______________________________

Year 20__ Date	Aircraft Type (Make, Model, Variant)	Aircraft Ident. Mark, Reg. No.	Departure & Arrival		Miles Flown	Flight No.	Number of		No. Instr. Appr.	Aircraft Category and Class				Pilot Function			
			From / Time	To / Time			TO (Day / Night)	LDG (Day / Night)		SE	ME			PIC (Solo, SPIC, PICUS)	Co-Pilot	Dual Received	Instructor/ Examiner/ Ck. Pilot
						Page Total											
						Amount Forward											
						Total to Date											

	Operational Conditions					Synthetic Flight				
	Cross-Country	Night	Actual Instrument	Instrument (Hood)		Time of Session	Type of Instruction	Type and Device No.	Total Duration of Flight	Remarks, Procedures, Maneuvers, Endorsements, PIC Name

I certify that the statements made by me on this form are true.

Pilot's Signature ______________________________

Year 20__ Date	Aircraft Type (Make, Model, Variant)	Aircraft Ident. Mark, Reg. No.	Departure & Arrival: From / Time	Departure & Arrival: To / Time	Miles Flown	Flight No.	Number of TO (Day / Night)	Number of LDG (Day / Night)	No. Instr. Appr.	Aircraft Category and Class: SE	ME			Pilot Function: PIC (Solo, SPIC, PICUS)	Co-Pilot	Dual Received	Instructor/ Examiner/ Ck. Pilot
						Page Total											
						Amount Forward											
						Total to Date											

	Operational Conditions					Synthetic Flight				
	Cross-Country	Night	Actual Instrument	Instrument (Hood)		Time of Session	Type of Instruction	Type and Device No.	Total Duration of Flight	Remarks, Procedures, Maneuvers, Endorsements, PIC Name

I certify that the statements made by me on this form are true.

Pilot's Signature ______________________

Year 20__ Date	Aircraft Type (Make, Model, Variant)	Aircraft Ident. Mark, Reg. No.	Departure & Arrival		Miles Flown	Flight No.	Number of		No. Instr. Appr.	Aircraft Category and Class				Pilot Function			
			From Time	To Time			TO Day Night	LDG Day Night		SE	ME			PIC (Solo, SPIC, PICUS)	Co-Pilot	Dual Received	Instructor/ Examiner/ Ck. Pilot
						Page Total											
						Amount Forward											
						Total to Date											

	Operational Conditions					Synthetic Flight				
	Cross-Country	Night	Actual Instrument	Instrument (Hood)		Time of Session	Type of Instruction	Type and Device No.	Total Duration of Flight	Remarks, Procedures, Maneuvers, Endorsements, PIC Name

I certify that the statements made by me on this form are true.

Pilot's Signature ______________________________

Year			Departure & Arrival				Number of			Aircraft Category and Class								Pilot Function							
20__ Date	Aircraft Type (Make, Model, Variant)	Aircraft Ident. Mark, Reg. No.	From Time	To Time	Miles Flown	Flight No.	TO Day Night	LDG Day Night	No. Instr. Appr.	SE		ME						PIC (Solo, SPIC, PICUS)		Co-Pilot		Dual Received		Instructor/ Examiner/ Ck. Pilot	
						Page Total																			
						Amount Forward																			
						Total to Date																			

	Operational Conditions					Synthetic Flight				
	Cross-Country	Night	Actual Instrument	Instrument (Hood)		Time of Session	Type of Instruction	Type and Device No.	Total Duration of Flight	Remarks, Procedures, Maneuvers, Endorsements, PIC Name

I certify that the statements made by me on this form are true.

Pilot's Signature ______________________________

Year 20__ Date	Aircraft Type (Make, Model, Variant)	Aircraft Ident. Mark, Reg. No.	Departure & Arrival		Miles Flown	Flight No.	Number of		No. Instr. Appr.	Aircraft Category and Class				Pilot Function			
			From Time	To Time			TO Day Night	LDG Day Night		SE	ME			PIC (Solo, SPIC, PICUS)	Co-Pilot	Dual Received	Instructor/ Examiner/ Ck. Pilot
							Page Total											
							Amount Forward											
							Total to Date											

		Operational Conditions										Synthetic Flight					
		Cross-Country		Night		Actual Instrument		Instrument (Hood)				Time of Session	Type of Instruction	Type and Device No.	Total Duration of Flight		Remarks, Procedures, Maneuvers, Endorsements, PIC Name

I certify that the statements made by me on this form are true.

Pilot's Signature ______________________________

Year 20__ Date	Aircraft Type (Make, Model, Variant)	Aircraft Ident. Mark, Reg. No.	Departure & Arrival		Miles Flown	Flight No.	Number of		No. Instr. Appr.	Aircraft Category and Class				Pilot Function			
			From Time	To Time			TO Day Night	LDG Day Night		SE	ME			PIC (Solo, SPIC, PICUS)	Co-Pilot	Dual Received	Instructor/ Examiner/ Ck. Pilot
						Page Total											
						Amount Forward											
						Total to Date											

		Operational Conditions										Synthetic Flight					
		Cross-Country		Night		Actual Instrument		Instrument (Hood)				Time of Session	Type of Instruction	Type and Device No.	Total Duration of Flight		Remarks, Procedures, Maneuvers, Endorsements, PIC Name

I certify that the statements made by me on this form are true.

Pilot's Signature ______________________

Year 20__ Date	Aircraft Type (Make, Model, Variant)	Aircraft Ident. Mark, Reg. No.	Departure & Arrival		Miles Flown	Flight No.	Number of		No. Instr. Appr.	Aircraft Category and Class				Pilot Function			
			From Time	To Time			TO Day Night	LDG Day Night		SE	ME			PIC (Solo, SPIC, PICUS)	Co-Pilot	Dual Received	Instructor/ Examiner/ Ck. Pilot
						Page Total											
						Amount Forward											
						Total to Date											

	Operational Conditions					Synthetic Flight				
	Cross-Country	Night	Actual Instrument	Instrument (Hood)		Time of Session	Type of Instruction	Type and Device No.	Total Duration of Flight	Remarks, Procedures, Maneuvers, Endorsements, PIC Name

I certify that the statements made by me on this form are true.

Pilot's Signature ______________________

			Departure & Arrival				Number of			**Aircraft Category and Class**								**Pilot Function**							
Year 20__ Date	Aircraft Type (Make, Model, Variant)	Aircraft Ident. Mark, Reg. No.	From Time	To Time	Miles Flown	Flight No.	TO Day Night	LDG Day Night	No. Instr. Appr.	SE		ME						PIC (Solo, SPIC, PICUS)		Co-Pilot		Dual Received		Instructor/ Examiner/ Ck. Pilot	
						Page Total																			
						Amount Forward																			
						Total to Date																			

		Operational Conditions										Synthetic Flight					
		Cross-Country		Night		Actual Instrument		Instrument (Hood)				Time of Session	Type of Instruction	Type and Device No.	Total Duration of Flight		Remarks, Procedures, Maneuvers, Endorsements, PIC Name

I certify that the statements made by me on this form are true.

Pilot's Signature ______________________________

Year 20__ Date	Aircraft Type (Make, Model, Variant)	Aircraft Ident. Mark, Reg. No.	Departure & Arrival: From / Time	To / Time	Miles Flown	Flight No.	Number of TO (Day / Night)	Number of LDG (Day / Night)	No. Instr. Appr.	Aircraft Category and Class: SE	ME			Pilot Function: PIC (Solo, SPIC, PICUS)	Co-Pilot	Dual Received	Instructor/ Examiner/ Ck. Pilot
						Page Total											
						Amount Forward											
						Total to Date											

	Operational Conditions					Synthetic Flight				
	Cross-Country	Night	Actual Instrument	Instrument (Hood)		Time of Session	Type of Instruction	Type and Device No.	Total Duration of Flight	Remarks, Procedures, Maneuvers, Endorsements, PIC Name

I certify that the statements made by me on this form are true.

Pilot's Signature ______________________

Year 20__ Date	Aircraft Type (Make, Model, Variant)	Aircraft Ident. Mark, Reg. No.	Departure & Arrival		Miles Flown	Flight No.	Number of		No. Instr. Appr.	Aircraft Category and Class				Pilot Function			
			From Time	To Time			TO Day Night	LDG Day Night		SE	ME			PIC (Solo, SPIC, PICUS)	Co-Pilot	Dual Received	Instructor/ Examiner/ Ck. Pilot
							Page Total										
							Amount Forward										
							Total to Date										

	Operational Conditions					Synthetic Flight				
	Cross-Country	Night	Actual Instrument	Instrument (Hood)		Time of Session	Type of Instruction	Type and Device No.	Total Duration of Flight	Remarks, Procedures, Maneuvers, Endorsements, PIC Name

I certify that the statements made by me on this form are true.

Pilot's Signature ______________________

Year 20__ Date	Aircraft Type (Make, Model, Variant)	Aircraft Ident. Mark, Reg. No.	Departure & Arrival				Miles Flown	Flight No.	Number of				No. Instr. Appr.	Aircraft Category and Class								Pilot Function							
			From	Time	To	Time			TO Day	TO Night	LDG Day	LDG Night		SE		ME						PIC (Solo, SPIC, PICUS)		Co-Pilot		Dual Received		Instructor/ Examiner/ Ck. Pilot	
								Page Total																					
								Amount Forward																					
								Total to Date																					

	Operational Conditions					Synthetic Flight				
	Cross-Country	Night	Actual Instrument	Instrument (Hood)		Time of Session	Type of Instruction	Type and Device No.	Total Duration of Flight	Remarks, Procedures, Maneuvers, Endorsements, PIC Name

I certify that the statements made by me on this form are true.

Pilot's Signature ______________________________

Year 20__ Date	Aircraft Type (Make, Model, Variant)	Aircraft Ident. Mark, Reg. No.	Departure & Arrival		Miles Flown	Flight No.	Number of		No. Instr. Appr.	Aircraft Category and Class								Pilot Function							
			From Time	To Time			TO Day Night	LDG Day Night		SE		ME						PIC (Solo, SPIC, PICUS)		Co-Pilot		Dual Received		Instructor/ Examiner/ Ck. Pilot	
						Page Total																			
						Amount Forward																			
						Total to Date																			

		Operational Conditions										Synthetic Flight					
		Cross-Country		Night		Actual Instrument		Instrument (Hood)				Time of Session	Type of Instruction	Type and Device No.	Total Duration of Flight		Remarks, Procedures, Maneuvers, Endorsements, PIC Name

I certify that the statements made by me on this form are true.

Pilot's Signature ______________________

Year 20__ Date	Aircraft Type (Make, Model, Variant)	Aircraft Ident. Mark, Reg. No.	Departure & Arrival: From Time	Departure & Arrival: To Time	Miles Flown	Flight No.	Number of TO: Day Night	Number of LDG: Day Night	No. Instr. Appr.	Aircraft Category and Class: SE		ME						Pilot Function: PIC (Solo, SPIC, PICUS)		Co-Pilot		Dual Received		Instructor/ Examiner/ Ck. Pilot	
						Page Total																			
						Amount Forward																			
						Total to Date																			

		Operational Conditions										Synthetic Flight					
		Cross-Country		Night		Actual Instrument		Instrument (Hood)				Time of Session	Type of Instruction	Type and Device No.	Total Duration of Flight		Remarks, Procedures, Maneuvers, Endorsements, PIC Name

I certify that the statements made by me on this form are true.

Pilot's Signature ______________________

Year 20__ Date	Aircraft Type (Make, Model, Variant)	Aircraft Ident. Mark, Reg. No.	**Departure & Arrival**		Miles Flown	Flight No.	Number of		No. Instr. Appr.	**Aircraft Category and Class**								**Pilot Function**							
			From Time	To Time			TO Day Night	LDG Day Night		SE		ME						PIC (Solo, SPIC, PICUS)		Co-Pilot		Dual Received		Instructor/ Examiner/ Ck. Pilot	
Page Total																									
Amount Forward																									
Total to Date																									

	Operational Conditions					Synthetic Flight				
	Cross-Country	Night	Actual Instrument	Instrument (Hood)		Time of Session	Type of Instruction	Type and Device No.	Total Duration of Flight	Remarks, Procedures, Maneuvers, Endorsements, PIC Name

I certify that the statements made by me on this form are true.

Pilot's Signature ______________________

Year 20__ Date	Aircraft Type (Make, Model, Variant)	Aircraft Ident. Mark, Reg. No.	Departure & Arrival: From / Time	To / Time	Miles Flown	Flight No.	Number of: TO (Day / Night)	LDG (Day / Night)	No. Instr. Appr.	Aircraft Category and Class: SE		ME						Pilot Function: PIC (Solo, SPIC, PICUS)		Co-Pilot		Dual Received		Instructor/ Examiner/ Ck. Pilot	
							Page Total																		
							Amount Forward																		
							Total to Date																		

	Operational Conditions					Synthetic Flight				
	Cross-Country	Night	Actual Instrument	Instrument (Hood)		Time of Session	Type of Instruction	Type and Device No.	Total Duration of Flight	Remarks, Procedures, Maneuvers, Endorsements, PIC Name
									I certify that the statements made by me on this form are true.
										Pilot's Signature ______________________

Year 20__ Date	Aircraft Type (Make, Model, Variant)	Aircraft Ident. Mark, Reg. No.	**Departure & Arrival**		Miles Flown	Flight No.	Number of		No. Instr. Appr.	**Aircraft Category and Class**								**Pilot Function**							
			From Time	To Time			TO Day Night	LDG Day Night		SE		ME						PIC (Solo, SPIC, PICUS)		Co-Pilot		Dual Received		Instructor/ Examiner/ Ck. Pilot	
						Page Total																			
						Amount Forward																			
						Total to Date																			

	Operational Conditions					Synthetic Flight				
	Cross-Country	Night	Actual Instrument	Instrument (Hood)		Time of Session	Type of Instruction	Type and Device No.	Total Duration of Flight	Remarks, Procedures, Maneuvers, Endorsements, PIC Name

I certify that the statements made by me on this form are true.

Pilot's Signature ______________________

Year 20__ Date	Aircraft Type (Make, Model, Variant)	Aircraft Ident. Mark, Reg. No.	**Departure & Arrival**		Miles Flown	Flight No.	Number of		No. Instr. Appr.	**Aircraft Category and Class**				**Pilot Function**			
			From Time	To Time			TO Day Night	LDG Day Night		SE	ME			PIC (Solo, SPIC, PICUS)	Co-Pilot	Dual Received	Instructor/ Examiner/ Ck. Pilot
						Page Total											
						Amount Forward											
						Total to Date											

	Operational Conditions					Synthetic Flight				
	Cross-Country	Night	Actual Instrument	Instrument (Hood)		Time of Session	Type of Instruction	Type and Device No.	Total Duration of Flight	Remarks, Procedures, Maneuvers, Endorsements, PIC Name

I certify that the statements made by me on this form are true.

Pilot's Signature ____________________

Year 20__ Date	Aircraft Type (Make, Model, Variant)	Aircraft Ident. Mark, Reg. No.	Departure & Arrival: From / Time	Departure & Arrival: To / Time	Miles Flown	Flight No.	Number of: TO (Day / Night)	Number of: LDG (Day / Night)	No. Instr. Appr.	Aircraft Category and Class: SE	Aircraft Category and Class: ME	Aircraft Category and Class	Aircraft Category and Class	Pilot Function: PIC (Solo, SPIC, PICUS)	Pilot Function: Co-Pilot	Pilot Function: Dual Received	Pilot Function: Instructor/ Examiner/ Ck. Pilot
						Page Total											
						Amount Forward											
						Total to Date											

		Operational Conditions										Synthetic Flight					
		Cross-Country		Night		Actual Instrument		Instrument (Hood)				Time of Session	Type of Instruction	Type and Device No.	Total Duration of Flight		Remarks, Procedures, Maneuvers, Endorsements, PIC Name

I certify that the statements made by me on this form are true.

Pilot's Signature ______________________

Year 20__ Date	Aircraft Type (Make, Model, Variant)	Aircraft Ident. Mark, Reg. No.	**Departure & Arrival**		Miles Flown	Flight No.	Number of		No. Instr. Appr.	**Aircraft Category and Class**				**Pilot Function**			
			From Time	To Time			TO Day Night	LDG Day Night		SE	ME			PIC (Solo, SPIC, PICUS)	Co-Pilot	Dual Received	Instructor/ Examiner/ Ck. Pilot
						Page Total											
						Amount Forward											
						Total to Date											

	Operational Conditions					Synthetic Flight				
	Cross-Country	Night	Actual Instrument	Instrument (Hood)		Time of Session	Type of Instruction	Type and Device No.	Total Duration of Flight	Remarks, Procedures, Maneuvers, Endorsements, PIC Name

I certify that the statements made by me on this form are true.

Pilot's Signature ______________________

Year 20__ Date	Aircraft Type (Make, Model, Variant)	Aircraft Ident. Mark, Reg. No.	Departure & Arrival		Miles Flown	Flight No.	Number of		No. Instr. Appr.	Aircraft Category and Class				Pilot Function			
			From / Time	To / Time			TO (Day / Night)	LDG (Day / Night)		SE	ME			PIC (Solo, SPIC, PICUS)	Co-Pilot	Dual Received	Instructor/ Examiner/ Ck. Pilot
						Page Total											
						Amount Forward											
						Total to Date											

	Operational Conditions					Synthetic Flight				
	Cross-Country	Night	Actual Instrument	Instrument (Hood)		Time of Session	Type of Instruction	Type and Device No.	Total Duration of Flight	Remarks, Procedures, Maneuvers, Endorsements, PIC Name

I certify that the statements made by me on this form are true.

Pilot's Signature ______________________________

Year 20__ Date	Aircraft Type (Make, Model, Variant)	Aircraft Ident. Mark, Reg. No.	Departure & Arrival		Miles Flown	Flight No.	Number of		No. Instr. Appr.	Aircraft Category and Class				Pilot Function			
			From Time	To Time			TO Day Night	LDG Day Night		SE	ME			PIC (Solo, SPIC, PICUS)	Co-Pilot	Dual Received	Instructor/ Examiner/ Ck. Pilot
						Page Total											
						Amount Forward											
						Total to Date											

	Operational Conditions					Synthetic Flight				
	Cross-Country	Night	Actual Instrument	Instrument (Hood)		Time of Session	Type of Instruction	Type and Device No.	Total Duration of Flight	Remarks, Procedures, Maneuvers, Endorsements, PIC Name

I certify that the statements made by me on this form are true.

Pilot's Signature ______________________________

Year 20__ Date	Aircraft Type (Make, Model, Variant)	Aircraft Ident. Mark, Reg. No.	**Departure & Arrival**		Miles Flown	Flight No.	Number of		No. Instr. Appr.	**Aircraft Category and Class**				**Pilot Function**			
			From Time	To Time			TO Day Night	LDG Day Night		SE	ME			PIC (Solo, SPIC, PICUS)	Co-Pilot	Dual Received	Instructor/ Examiner/ Ck. Pilot
						Page Total											
						Amount Forward											
						Total to Date											

	Operational Conditions					Synthetic Flight				
	Cross-Country	Night	Actual Instrument	Instrument (Hood)		Time of Session	Type of Instruction	Type and Device No.	Total Duration of Flight	Remarks, Procedures, Maneuvers, Endorsements, PIC Name

I certify that the statements made by me on this form are true.

Pilot's Signature ______________________________

Year 20__ Date	Aircraft Type (Make, Model, Variant)	Aircraft Ident. Mark, Reg. No.	Departure & Arrival		Miles Flown	Flight No.	Number of		No. Instr. Appr.	Aircraft Category and Class								Pilot Function							
			From Time	To Time			TO Day Night	LDG Day Night		SE		ME						PIC (Solo, SPIC, PICUS)		Co-Pilot		Dual Received		Instructor/ Examiner/ Ck. Pilot	
							Page Total																		
							Amount Forward																		
							Total to Date																		

	Operational Conditions					Synthetic Flight				
	Cross-Country	Night	Actual Instrument	Instrument (Hood)		Time of Session	Type of Instruction	Type and Device No.	Total Duration of Flight	Remarks, Procedures, Maneuvers, Endorsements, PIC Name

I certify that the statements made by me on this form are true.

Pilot's Signature ______________________________

			Departure & Arrival				Number of			Aircraft Category and Class				Pilot Function			
Year 20__ Date	Aircraft Type (Make, Model, Variant)	Aircraft Ident. Mark, Reg. No.	From Time	To Time	Miles Flown	Flight No.	TO Day Night	LDG Day Night	No. Instr. Appr.	SE	ME			PIC (Solo, SPIC, PICUS)	Co-Pilot	Dual Received	Instructor/ Examiner/ Ck. Pilot
						Page Total											
						Amount Forward											
						Total to Date											

		Operational Conditions										Synthetic Flight					
		Cross-Country		Night		Actual Instrument		Instrument (Hood)				Time of Session	Type of Instruction	Type and Device No.	Total Duration of Flight		Remarks, Procedures, Maneuvers, Endorsements, PIC Name

I certify that the statements made by me on this form are true.

Pilot's Signature ______________________________

Year 20__ Date	Aircraft Type (Make, Model, Variant)	Aircraft Ident. Mark, Reg. No.	Departure & Arrival		Miles Flown	Flight No.	Number of		No. Instr. Appr.	Aircraft Category and Class				Pilot Function			
			From Time	To Time			TO Day Night	LDG Day Night		SE	ME			PIC (Solo, SPIC, PICUS)	Co-Pilot	Dual Received	Instructor/ Examiner/ Ck. Pilot
						Page Total											
						Amount Forward											
						Total to Date											

		Operational Conditions										Synthetic Flight					
		Cross-Country		Night		Actual Instrument		Instrument (Hood)				Time of Session	Type of Instruction	Type and Device No.	Total Duration of Flight		Remarks, Procedures, Maneuvers, Endorsements, PIC Name

I certify that the statements made by me on this form are true.

Pilot's Signature ______________________________

Year 20__ Date	Aircraft Type (Make, Model, Variant)	Aircraft Ident. Mark, Reg. No.	Departure & Arrival: From / Time	Departure & Arrival: To / Time	Miles Flown	Flight No.	Number of: TO (Day / Night)	Number of: LDG (Day / Night)	No. Instr. Appr.	Aircraft Category and Class: SE	ME			Pilot Function: PIC (Solo, SPIC, PICUS)	Co-Pilot	Dual Received	Instructor/ Examiner/ Ck. Pilot
						Page Total											
						Amount Forward											
						Total to Date											

		Operational Conditions										Synthetic Flight			Total Duration of Flight		Remarks, Procedures, Maneuvers, Endorsements, PIC Name
		Cross-Country		Night		Actual Instrument		Instrument (Hood)				Time of Session	Type of Instruction	Type and Device No.			

I certify that the statements made by me on this form are true.

Pilot's Signature ______________________________

Year 20__ Date	Aircraft Type (Make, Model, Variant)	Aircraft Ident. Mark, Reg. No.	**Departure & Arrival**		Miles Flown	Flight No.	Number of		No. Instr. Appr.	**Aircraft Category and Class**								**Pilot Function**							
			From Time	To Time			TO Day Night	LDG Day Night		SE		ME						PIC (Solo, SPIC, PICUS)		Co-Pilot		Dual Received		Instructor/ Examiner/ Ck. Pilot	
						Page Total																			
						Amount Forward																			
						Total to Date																			

		Operational Conditions										Synthetic Flight					
		Cross-Country		Night		Actual Instrument		Instrument (Hood)				Time of Session	Type of Instruction	Type and Device No.	Total Duration of Flight		Remarks, Procedures, Maneuvers, Endorsements, PIC Name

I certify that the statements made by me on this form are true.

Pilot's Signature ______________________________

Year 20__	Aircraft Type	Aircraft	Departure & Arrival		Miles	Flight	Number of		No.	Aircraft Category and Class								Pilot Function							
Date	(Make, Model, Variant)	Ident. Mark, Reg. No.	From Time	To Time	Flown	No.	TO Day Night	LDG Day Night	Instr. Appr.	SE		ME						PIC (Solo, SPIC, PICUS)		Co-Pilot		Dual Received		Instructor/ Examiner/ Ck. Pilot	
						Page Total																			
						Amount Forward																			
						Total to Date																			

	Operational Conditions					Synthetic Flight				
	Cross-Country	Night	Actual Instrument	Instrument (Hood)		Time of Session	Type of Instruction	Type and Device No.	Total Duration of Flight	Remarks, Procedures, Maneuvers, Endorsements, PIC Name

I certify that the statements made by me on this form are true.

Pilot's Signature ____________________

Year 20__ Date	Aircraft Type (Make, Model, Variant)	Aircraft Ident. Mark, Reg. No.	Departure & Arrival: From / Time	Departure & Arrival: To / Time	Miles Flown	Flight No.	Number of: TO (Day / Night)	Number of: LDG (Day / Night)	No. Instr. Appr.	Aircraft Category and Class: SE		ME						Pilot Function: PIC (Solo, SPIC, PICUS)		Co-Pilot		Dual Received		Instructor/ Examiner/ Ck. Pilot	
						Page Total																			
						Amount Forward																			
						Total to Date																			

		Operational Conditions										Synthetic Flight					
		Cross-Country		Night		Actual Instrument		Instrument (Hood)				Time of Session	Type of Instruction	Type and Device No.	Total Duration of Flight		Remarks, Procedures, Maneuvers, Endorsements, PIC Name

I certify that the statements made by me on this form are true.

Pilot's Signature ______________________________

Year 20__ Date	Aircraft Type (Make, Model, Variant)	Aircraft Ident. Mark, Reg. No.	Departure & Arrival		Miles Flown	Flight No.	Number of		No. Instr. Appr.	Aircraft Category and Class								Pilot Function							
			From Time	To Time			TO Day Night	LDG Day Night		SE		ME						PIC (Solo, SPIC, PICUS)		Co-Pilot		Dual Received		Instructor/ Examiner/ Ck. Pilot	
						Page Total																			
						Amount Forward																			
						Total to Date																			

	Operational Conditions					Synthetic Flight				
	Cross-Country	Night	Actual Instrument	Instrument (Hood)		Time of Session	Type of Instruction	Type and Device No.	Total Duration of Flight	Remarks, Procedures, Maneuvers, Endorsements, PIC Name

I certify that the statements made by me on this form are true.

Pilot's Signature ______________________________

Year 20__ Date	Aircraft Type (Make, Model, Variant)	Aircraft Ident. Mark, Reg. No.	Departure & Arrival		Miles Flown	Flight No.	Number of		No. Instr. Appr.	Aircraft Category and Class				Pilot Function			
			From Time	To Time			TO Day Night	LDG Day Night		SE	ME			PIC (Solo, SPIC, PICUS)	Co-Pilot	Dual Received	Instructor/ Examiner/ Ck. Pilot
						Page Total											
						Amount Forward											
						Total to Date											

		Operational Conditions										Synthetic Flight					
		Cross-Country		Night		Actual Instrument		Instrument (Hood)				Time of Session	Type of Instruction	Type and Device No.	Total Duration of Flight		Remarks, Procedures, Maneuvers, Endorsements, PIC Name

I certify that the statements made by me on this form are true.

Pilot's Signature ______________________________

Year 20__ Date	Aircraft Type (Make, Model, Variant)	Aircraft Ident. Mark, Reg. No.	Departure & Arrival		Miles Flown	Flight No.	Number of		No. Instr. Appr.	Aircraft Category and Class				Pilot Function			
			From Time	To Time			TO Day Night	LDG Day Night		SE	ME			PIC (Solo, SPIC, PICUS)	Co-Pilot	Dual Received	Instructor/ Examiner/ Ck. Pilot
						Page Total											
						Amount Forward											
						Total to Date											

	Operational Conditions					Synthetic Flight				
	Cross-Country	Night	Actual Instrument	Instrument (Hood)		Time of Session	Type of Instruction	Type and Device No.	Total Duration of Flight	Remarks, Procedures, Maneuvers, Endorsements, PIC Name

I certify that the statements made by me on this form are true.

Pilot's Signature ____________________

Year 20__ Date	Aircraft Type (Make, Model, Variant)	Aircraft Ident. Mark, Reg. No.	Departure & Arrival		Miles Flown	Flight No.	Number of		No. Instr. Appr.	Aircraft Category and Class				Pilot Function			
			From Time	To Time			TO Day Night	LDG Day Night		SE	ME			PIC (Solo, SPIC, PICUS)	Co-Pilot	Dual Received	Instructor/ Examiner/ Ck. Pilot
						Page Total											
						Amount Forward											
						Total to Date											

	Operational Conditions					Synthetic Flight				
	Cross-Country	Night	Actual Instrument	Instrument (Hood)		Time of Session	Type of Instruction	Type and Device No.	Total Duration of Flight	Remarks, Procedures, Maneuvers, Endorsements, PIC Name

I certify that the statements made by me on this form are true.

Pilot's Signature ____________________

Year 20__ Date	Aircraft Type (Make, Model, Variant)	Aircraft Ident. Mark, Reg. No.	Departure & Arrival		Miles Flown	Flight No.	Number of		No. Instr. Appr.	Aircraft Category and Class								Pilot Function							
			From Time	To Time			TO Day Night	LDG Day Night		SE		ME						PIC (Solo, SPIC, PICUS)		Co-Pilot		Dual Received		Instructor/ Examiner/ Ck. Pilot	
						Page Total																			
						Amount Forward																			
						Total to Date																			

	Operational Conditions					Synthetic Flight				
	Cross-Country	Night	Actual Instrument	Instrument (Hood)		Time of Session	Type of Instruction	Type and Device No.	Total Duration of Flight	Remarks, Procedures, Maneuvers, Endorsements, PIC Name

I certify that the statements made by me on this form are true.

Pilot's Signature ______________________________

Year 20__ Date	Aircraft Type (Make, Model, Variant)	Aircraft Ident. Mark, Reg. No.	Departure & Arrival		Miles Flown	Flight No.	Number of		No. Instr. Appr.	Aircraft Category and Class								Pilot Function							
			From Time	To Time			TO Day Night	LDG Day Night		SE		ME						PIC (Solo, SPIC, PICUS)		Co-Pilot		Dual Received		Instructor/ Examiner/ Ck. Pilot	
						Page Total																			
						Amount Forward																			
						Total to Date																			

		Operational Conditions										Synthetic Flight					
		Cross-Country		Night		Actual Instrument		Instrument (Hood)				Time of Session	Type of Instruction	Type and Device No.	Total Duration of Flight		Remarks, Procedures, Maneuvers, Endorsements, PIC Name

I certify that the statements made by me on this form are true.

Pilot's Signature ______________________________

Year 20__ Date	Aircraft Type (Make, Model, Variant)	Aircraft Ident. Mark, Reg. No.	Departure & Arrival		Miles Flown	Flight No.	Number of		No. Instr. Appr.	Aircraft Category and Class				Pilot Function			
			From Time	To Time			TO Day Night	LDG Day Night		SE	ME			PIC (Solo, SPIC, PICUS)	Co-Pilot	Dual Received	Instructor/ Examiner/ Ck. Pilot
						Page Total											
						Amount Forward											
						Total to Date											

	Operational Conditions					Synthetic Flight				
	Cross-Country	Night	Actual Instrument	Instrument (Hood)		Time of Session	Type of Instruction	Type and Device No.	Total Duration of Flight	Remarks, Procedures, Maneuvers, Endorsements, PIC Name

I certify that the statements made by me on this form are true.

Pilot's Signature ____________________

Year 20__ Date	Aircraft Type (Make, Model, Variant)	Aircraft Ident. Mark, Reg. No.	Departure & Arrival		Miles Flown	Flight No.	Number of		No. Instr. Appr.	Aircraft Category and Class								Pilot Function							
			From Time	To Time			TO Day Night	LDG Day Night		SE		ME						PIC (Solo, SPIC, PICUS)		Co-Pilot		Dual Received		Instructor/ Examiner/ Ck. Pilot	
						Page Total																			
						Amount Forward																			
						Total to Date																			

		Operational Conditions										Synthetic Flight			Total Duration of Flight		Remarks, Procedures, Maneuvers, Endorsements, PIC Name
		Cross-Country		Night		Actual Instrument		Instrument (Hood)				Time of Session	Type of Instruction	Type and Device No.			

I certify that the statements made by me on this form are true.

Pilot's Signature ______________________________

Year 20__ Date	Aircraft Type (Make, Model, Variant)	Aircraft Ident. Mark, Reg. No.	Departure & Arrival		Miles Flown	Flight No.	Number of		No. Instr. Appr.	Aircraft Category and Class				Pilot Function			
			From Time	To Time			TO Day Night	LDG Day Night		SE	ME			PIC (Solo, SPIC, PICUS)	Co-Pilot	Dual Received	Instructor/ Examiner/ Ck. Pilot
							Page Total										
							Amount Forward										
							Total to Date										

		Operational Conditions										Synthetic Flight					
		Cross-Country		Night		Actual Instrument		Instrument (Hood)				Time of Session	Type of Instruction	Type and Device No.	Total Duration of Flight		Remarks, Procedures, Maneuvers, Endorsements, PIC Name

I certify that the statements made by me on this form are true.

Pilot's Signature ______________________________

Year 20__ Date	Aircraft Type (Make, Model, Variant)	Aircraft Ident. Mark, Reg. No.	**Departure & Arrival**		Miles Flown	Flight No.	Number of		No. Instr. Appr.	**Aircraft Category and Class**								**Pilot Function**							
			From Time	To Time			TO Day Night	LDG Day Night		SE		ME						PIC (Solo, SPIC, PICUS)		Co-Pilot		Dual Received		Instructor/ Examiner/ Ck. Pilot	
								Page Total																	
								Amount Forward																	
								Total to Date																	

		Operational Conditions										Synthetic Flight					
		Cross-Country		Night		Actual Instrument		Instrument (Hood)				Time of Session	Type of Instruction	Type and Device No.	Total Duration of Flight		Remarks, Procedures, Maneuvers, Endorsements, PIC Name

I certify that the statements made by me on this form are true.

Pilot's Signature ____________________

Year 20__ Date	Aircraft Type (Make, Model, Variant)	Aircraft Ident. Mark, Reg. No.	Departure & Arrival		Miles Flown	Flight No.	Number of		No. Instr. Appr.	Aircraft Category and Class								Pilot Function							
			From Time	To Time			TO Day Night	LDG Day Night		SE		ME						PIC (Solo, SPIC, PICUS)		Co-Pilot		Dual Received		Instructor/ Examiner/ Ck. Pilot	
						Page Total																			
						Amount Forward																			
						Total to Date																			

		Operational Conditions										Synthetic Flight					
		Cross-Country		Night		Actual Instrument		Instrument (Hood)				Time of Session	Type of Instruction	Type and Device No.	Total Duration of Flight		Remarks, Procedures, Maneuvers, Endorsements, PIC Name

I certify that the statements made by me on this form are true.

Pilot's Signature ______________________________

Year 20__ Date	Aircraft Type (Make, Model, Variant)	Aircraft Ident. Mark, Reg. No.	Departure & Arrival		Miles Flown	Flight No.	Number of		No. Instr. Appr.	Aircraft Category and Class								Pilot Function							
			From Time	To Time			TO Day Night	LDG Day Night		SE		ME						PIC (Solo, SPIC, PICUS)		Co-Pilot		Dual Received		Instructor/ Examiner/ Ck. Pilot	
						Page Total																			
						Amount Forward																			
						Total to Date																			

		Operational Conditions										Synthetic Flight					
		Cross-Country		Night		Actual Instrument		Instrument (Hood)				Time of Session	Type of Instruction	Type and Device No.	Total Duration of Flight		Remarks, Procedures, Maneuvers, Endorsements, PIC Name

I certify that the statements made by me on this form are true.

Pilot's Signature ______________________________

Year 20__ Date	Aircraft Type (Make, Model, Variant)	Aircraft Ident. Mark, Reg. No.	**Departure & Arrival**		Miles Flown	Flight No.	Number of		No. Instr. Appr.	**Aircraft Category and Class**				**Pilot Function**			
			From Time	To Time			TO Day Night	LDG Day Night		SE	ME			PIC (Solo, SPIC, PICUS)	Co-Pilot	Dual Received	Instructor/ Examiner/ Ck. Pilot
						Page Total											
						Amount Forward											
						Total to Date											

	Operational Conditions					Synthetic Flight				
	Cross-Country	Night	Actual Instrument	Instrument (Hood)		Time of Session	Type of Instruction	Type and Device No.	Total Duration of Flight	Remarks, Procedures, Maneuvers, Endorsements, PIC Name

I certify that the statements made by me on this form are true.

Pilot's Signature ______________________________

Year 20__ Date	Aircraft Type (Make, Model, Variant)	Aircraft Ident. Mark, Reg. No.	Departure & Arrival		Miles Flown	Flight No.	Number of		No. Instr. Appr.	Aircraft Category and Class				Pilot Function			
			From Time	To Time			TO Day Night	LDG Day Night		SE	ME			PIC (Solo, SPIC, PICUS)	Co-Pilot	Dual Received	Instructor/ Examiner/ Ck. Pilot
							Page Total										
							Amount Forward										
							Total to Date										

		Operational Conditions										Synthetic Flight					
		Cross-Country		Night		Actual Instrument		Instrument (Hood)				Time of Session	Type of Instruction	Type and Device No.	Total Duration of Flight		Remarks, Procedures, Maneuvers, Endorsements, PIC Name

I certify that the statements made by me on this form are true.

Pilot's Signature ______________________________

Year 20__ Date	Aircraft Type (Make, Model, Variant)	Aircraft Ident. Mark, Reg. No.	Departure & Arrival: From Time	Departure & Arrival: To Time	Miles Flown	Flight No.	Number of TO (Day / Night)	Number of LDG (Day / Night)	No. Instr. Appr.	Aircraft Category and Class: SE		ME						Pilot Function: PIC (Solo, SPIC, PICUS)		Co-Pilot		Dual Received		Instructor/ Examiner/ Ck. Pilot	
						Page Total																			
						Amount Forward																			
						Total to Date																			

		Operational Conditions										Synthetic Flight					
		Cross-Country		Night		Actual Instrument		Instrument (Hood)				Time of Session	Type of Instruction	Type and Device No.	Total Duration of Flight		Remarks, Procedures, Maneuvers, Endorsements, PIC Name

I certify that the statements made by me on this form are true.

Pilot's Signature ____________________

Year 20__ Date	Aircraft Type (Make, Model, Variant)	Aircraft Ident. Mark, Reg. No.	**Departure & Arrival**		Miles Flown	Flight No.	Number of		No. Instr. Appr.	**Aircraft Category and Class**								**Pilot Function**							
			From Time	To Time			TO Day Night	LDG Day Night		SE		ME						PIC (Solo, SPIC, PICUS)		Co-Pilot		Dual Received		Instructor/ Examiner/ Ck. Pilot	
						Page Total																			
						Amount Forward																			
						Total to Date																			

		Operational Conditions										Synthetic Flight			Total Duration of Flight		Remarks, Procedures, Maneuvers, Endorsements, PIC Name
		Cross-Country		Night		Actual Instrument		Instrument (Hood)				Time of Session	Type of Instruction	Type and Device No.			

I certify that the statements made by me on this form are true.

Pilot's Signature ______________________________

Year 20__ Date	Aircraft Type (Make, Model, Variant)	Aircraft Ident. Mark, Reg. No.	**Departure & Arrival**		Miles Flown	Flight No.	Number of		No. Instr. Appr.	**Aircraft Category and Class**								**Pilot Function**							
			From / Time	To / Time			TO Day / Night	LDG Day / Night		SE		ME						PIC (Solo, SPIC, PICUS)		Co-Pilot		Dual Received		Instructor/ Examiner/ Ck. Pilot	
						Page Total																			
						Amount Forward																			
						Total to Date																			

	Operational Conditions					Synthetic Flight				
	Cross-Country	Night	Actual Instrument	Instrument (Hood)		Time of Session	Type of Instruction	Type and Device No.	Total Duration of Flight	Remarks, Procedures, Maneuvers, Endorsements, PIC Name

I certify that the statements made by me on this form are true.

Pilot's Signature ______________________________

Year 20__ Date	Aircraft Type (Make, Model, Variant)	Aircraft Ident. Mark, Reg. No.	**Departure & Arrival**		Miles Flown	Flight No.	Number of		No. Instr. Appr.	**Aircraft Category and Class**								**Pilot Function**							
			From Time	To Time			TO Day Night	LDG Day Night		SE		ME						PIC (Solo, SPIC, PICUS)		Co-Pilot		Dual Received		Instructor/ Examiner/ Ck. Pilot	
						Page Total																			
						Amount Forward																			
						Total to Date																			

		Operational Conditions										Synthetic Flight					
		Cross-Country		Night		Actual Instrument		Instrument (Hood)				Time of Session	Type of Instruction	Type and Device No.	Total Duration of Flight		Remarks, Procedures, Maneuvers, Endorsements, PIC Name

I certify that the statements made by me on this form are true.

Pilot's Signature ______________________________

Year 20__ Date	Aircraft Type (Make, Model, Variant)	Aircraft Ident. Mark, Reg. No.	**Departure & Arrival**		Miles Flown	Flight No.	Number of		No. Instr. Appr.	**Aircraft Category and Class**								**Pilot Function**							
			From / Time	To / Time			TO / Day / Night	LDG / Day / Night		SE		ME						PIC (Solo, SPIC, PICUS)		Co-Pilot		Dual Received		Instructor/ Examiner/ Ck. Pilot	
						Page Total																			
						Amount Forward																			
						Total to Date																			

		Operational Conditions										Synthetic Flight			Total Duration of Flight		Remarks, Procedures, Maneuvers, Endorsements, PIC Name
		Cross-Country		Night		Actual Instrument		Instrument (Hood)				Time of Session	Type of Instruction	Type and Device No.			

I certify that the statements made by me on this form are true.

Pilot's Signature ______________________

Year 20__ Date	Aircraft Type (Make, Model, Variant)	Aircraft Ident. Mark, Reg. No.	Departure & Arrival		Miles Flown	Flight No.	Number of		No. Instr. Appr.	Aircraft Category and Class				Pilot Function			
			From Time	To Time			TO Day Night	LDG Day Night		SE	ME			PIC (Solo, SPIC, PICUS)	Co-Pilot	Dual Received	Instructor/ Examiner/ Ck. Pilot
						Page Total											
						Amount Forward											
						Total to Date											

	Operational Conditions					Synthetic Flight				
	Cross-Country	Night	Actual Instrument	Instrument (Hood)		Time of Session	Type of Instruction	Type and Device No.	Total Duration of Flight	Remarks, Procedures, Maneuvers, Endorsements, PIC Name

I certify that the statements made by me on this form are true.

Pilot's Signature ____________________

Year 20__ Date	Aircraft Type (Make, Model, Variant)	Aircraft Ident. Mark, Reg. No.	**Departure & Arrival**		Miles Flown	Flight No.	Number of		No. Instr. Appr.	**Aircraft Category and Class**								**Pilot Function**							
			From Time	To Time			TO Day / Night	LDG Day / Night		SE		ME						PIC (Solo, SPIC, PICUS)		Co-Pilot		Dual Received		Instructor/ Examiner/ Ck. Pilot	
						Page Total																			
						Amount Forward																			
						Total to Date																			

	Operational Conditions					Synthetic Flight				
	Cross-Country	Night	Actual Instrument	Instrument (Hood)		Time of Session	Type of Instruction	Type and Device No.	Total Duration of Flight	Remarks, Procedures, Maneuvers, Endorsements, PIC Name

I certify that the statements made by me on this form are true.

Pilot's Signature ______________________________

Year 20__ Date	Aircraft Type (Make, Model, Variant)	Aircraft Ident. Mark, Reg. No.	Departure & Arrival: From / Time	Departure & Arrival: To / Time	Miles Flown	Flight No.	Number of: TO (Day / Night)	Number of: LDG (Day / Night)	No. Instr. Appr.	Aircraft Category and Class: SE	ME			Pilot Function: PIC (Solo, SPIC, PICUS)	Co-Pilot	Dual Received	Instructor/ Examiner/ Ck. Pilot
						Page Total											
						Amount Forward											
						Total to Date											

		Operational Conditions										Synthetic Flight					
		Cross-Country		Night		Actual Instrument		Instrument (Hood)				Time of Session	Type of Instruction	Type and Device No.	Total Duration of Flight		Remarks, Procedures, Maneuvers, Endorsements, PIC Name

I certify that the statements made by me on this form are true.

Pilot's Signature ______________________________

Year 20__ Date	Aircraft Type (Make, Model, Variant)	Aircraft Ident. Mark, Reg. No.	Departure & Arrival		Miles Flown	Flight No.	Number of		No. Instr. Appr.	Aircraft Category and Class				Pilot Function			
			From Time	To Time			TO Day Night	LDG Day Night		SE	ME			PIC (Solo, SPIC, PICUS)	Co-Pilot	Dual Received	Instructor/ Examiner/ Ck. Pilot
						Page Total											
						Amount Forward											
						Total to Date											

	Operational Conditions					Synthetic Flight				
	Cross-Country	Night	Actual Instrument	Instrument (Hood)		Time of Session	Type of Instruction	Type and Device No.	Total Duration of Flight	Remarks, Procedures, Maneuvers, Endorsements, PIC Name

I certify that the statements made by me on this form are true.

Pilot's Signature ______________________________

Year 20__ Date	Aircraft Type (Make, Model, Variant)	Aircraft Ident. Mark, Reg. No.	Departure & Arrival		Miles Flown	Flight No.	Number of		No. Instr. Appr.	Aircraft Category and Class								Pilot Function							
			From Time	To Time			TO Day Night	LDG Day Night		SE		ME						PIC (Solo, SPIC, PICUS)		Co-Pilot		Dual Received		Instructor/ Examiner/ Ck. Pilot	
						Page Total																			
						Amount Forward																			
						Total to Date																			

	Operational Conditions					Synthetic Flight				
	Cross-Country	Night	Actual Instrument	Instrument (Hood)		Time of Session	Type of Instruction	Type and Device No.	Total Duration of Flight	Remarks, Procedures, Maneuvers, Endorsements, PIC Name

I certify that the statements made by me on this form are true.

Pilot's Signature ________________________________

Year 20__ Date	Aircraft Type (Make, Model, Variant)	Aircraft Ident. Mark, Reg. No.	Departure & Arrival		Miles Flown	Flight No.	Number of		No. Instr. Appr.	Aircraft Category and Class				Pilot Function			
			From Time	To Time			TO Day Night	LDG Day Night		SE	ME			PIC (Solo, SPIC, PICUS)	Co-Pilot	Dual Received	Instructor/ Examiner/ Ck. Pilot
						Page Total											
						Amount Forward											
						Total to Date											

		Operational Conditions										Synthetic Flight					
		Cross-Country		Night		Actual Instrument		Instrument (Hood)				Time of Session	Type of Instruction	Type and Device No.	Total Duration of Flight		Remarks, Procedures, Maneuvers, Endorsements, PIC Name

I certify that the statements made by me on this form are true.

Pilot's Signature ______________________________

Year 20__ Date	Aircraft Type (Make, Model, Variant)	Aircraft Ident. Mark, Reg. No.	**Departure & Arrival**		Miles Flown	Flight No.	Number of		No. Instr. Appr.	**Aircraft Category and Class**								**Pilot Function**							
			From Time	To Time			TO Day Night	LDG Day Night		SE		ME						PIC (Solo, SPIC, PICUS)		Co-Pilot		Dual Received		Instructor/ Examiner/ Ck. Pilot	
						Page Total																			
						Amount Forward																			
						Total to Date																			

	Operational Conditions					Synthetic Flight				
	Cross-Country	Night	Actual Instrument	Instrument (Hood)		Time of Session	Type of Instruction	Type and Device No.	Total Duration of Flight	Remarks, Procedures, Maneuvers, Endorsements, PIC Name

I certify that the statements made by me on this form are true.

Pilot's Signature ______________________________

Year 20__ Date	Aircraft Type (Make, Model, Variant)	Aircraft Ident. Mark, Reg. No.	**Departure & Arrival**		Miles Flown	Flight No.	Number of		No. Instr. Appr.	**Aircraft Category and Class**				**Pilot Function**			
			From Time	To Time			TO Day Night	LDG Day Night		SE	ME			PIC (Solo, SPIC, PICUS)	Co-Pilot	Dual Received	Instructor/ Examiner/ Ck. Pilot
						Page Total											
						Amount Forward											
						Total to Date											

		Operational Conditions										Synthetic Flight					
		Cross-Country		Night		Actual Instrument		Instrument (Hood)				Time of Session	Type of Instruction	Type and Device No.	Total Duration of Flight		Remarks, Procedures, Maneuvers, Endorsements, PIC Name

I certify that the statements made by me on this form are true.

Pilot's Signature ______________________

Year 20__ Date	Aircraft Type (Make, Model, Variant)	Aircraft Ident. Mark, Reg. No.	**Departure & Arrival**		Miles Flown	Flight No.	Number of		No. Instr. Appr.	**Aircraft Category and Class**								**Pilot Function**							
			From Time	To Time			TO Day Night	LDG Day Night		SE		ME						PIC (Solo, SPIC, PICUS)		Co-Pilot		Dual Received		Instructor/ Examiner/ Ck. Pilot	
						Page Total																			
						Amount Forward																			
						Total to Date																			

	Operational Conditions					Synthetic Flight				
	Cross-Country	Night	Actual Instrument	Instrument (Hood)		Time of Session	Type of Instruction	Type and Device No.	Total Duration of Flight	Remarks, Procedures, Maneuvers, Endorsements, PIC Name

I certify that the statements made by me on this form are true.

Pilot's Signature ______________________

Year 20__ Date	Aircraft Type (Make, Model, Variant)	Aircraft Ident. Mark, Reg. No.	**Departure & Arrival**		Miles Flown	Flight No.	Number of		No. Instr. Appr.	**Aircraft Category and Class**								**Pilot Function**							
			From Time	To Time			TO Day Night	LDG Day Night		SE		ME						PIC (Solo, SPIC, PICUS)		Co-Pilot		Dual Received		Instructor/ Examiner/ Ck. Pilot	
						Page Total																			
						Amount Forward																			
						Total to Date																			

	Operational Conditions					Synthetic Flight				
	Cross-Country	Night	Actual Instrument	Instrument (Hood)		Time of Session	Type of Instruction	Type and Device No.	Total Duration of Flight	Remarks, Procedures, Maneuvers, Endorsements, PIC Name

I certify that the statements made by me on this form are true.

Pilot's Signature ______________________________

Year 20__	Aircraft Type	Aircraft	Departure & Arrival		Miles	Flight	Number of		No.	Aircraft Category and Class				Pilot Function			
Date	(Make, Model, Variant)	Ident. Mark, Reg. No.	From Time	To Time	Flown	No.	TO Day Night	LDG Day Night	Instr. Appr.	SE	ME			PIC (Solo, SPIC, PICUS)	Co-Pilot	Dual Received	Instructor/ Examiner/ Ck. Pilot
						Page Total											
						Amount Forward											
						Total to Date											

		Operational Conditions										Synthetic Flight					
		Cross-Country		Night		Actual Instrument		Instrument (Hood)				Time of Session	Type of Instruction	Type and Device No.	Total Duration of Flight		Remarks, Procedures, Maneuvers, Endorsements, PIC Name

I certify that the statements made by me on this form are true.

Pilot's Signature ______________________________

Year 20__ Date	Aircraft Type (Make, Model, Variant)	Aircraft Ident. Mark, Reg. No.	**Departure & Arrival**		Miles Flown	Flight No.	Number of		No. Instr. Appr.	**Aircraft Category and Class**								**Pilot Function**							
			From Time	To Time			TO Day Night	LDG Day Night		SE		ME						PIC (Solo, SPIC, PICUS)		Co-Pilot		Dual Received		Instructor/ Examiner/ Ck. Pilot	
						Page Total																			
						Amount Forward																			
						Total to Date																			

		Operational Conditions										Synthetic Flight					
		Cross-Country		Night		Actual Instrument		Instrument (Hood)				Time of Session	Type of Instruction	Type and Device No.	Total Duration of Flight		Remarks, Procedures, Maneuvers, Endorsements, PIC Name

I certify that the statements made by me on this form are true.

Pilot's Signature ______________________________

Year 20__ Date	Aircraft Type (Make, Model, Variant)	Aircraft Ident. Mark, Reg. No.	Departure & Arrival		Miles Flown	Flight No.	Number of		No. Instr. Appr.	Aircraft Category and Class				Pilot Function			
			From Time	To Time			TO Day Night	LDG Day Night		SE	ME			PIC (Solo, SPIC, PICUS)	Co-Pilot	Dual Received	Instructor/ Examiner/ Ck. Pilot
							Page Total										
							Amount Forward										
							Total to Date										

	Operational Conditions					Synthetic Flight				
	Cross-Country	Night	Actual Instrument	Instrument (Hood)		Time of Session	Type of Instruction	Type and Device No.	Total Duration of Flight	Remarks, Procedures, Maneuvers, Endorsements, PIC Name

I certify that the statements made by me on this form are true.

Pilot's Signature ______________________

Year 20__ Date	Aircraft Type (Make, Model, Variant)	Aircraft Ident. Mark, Reg. No.	**Departure & Arrival**		Miles Flown	Flight No.	Number of		No. Instr. Appr.	**Aircraft Category and Class**								**Pilot Function**							
			From Time	To Time			TO Day Night	LDG Day Night		SE		ME						PIC (Solo, SPIC, PICUS)		Co-Pilot		Dual Received		Instructor/ Examiner/ Ck. Pilot	
						Page Total																			
						Amount Forward																			
						Total to Date																			

	Operational Conditions					Synthetic Flight				
	Cross-Country	Night	Actual Instrument	Instrument (Hood)		Time of Session	Type of Instruction	Type and Device No.	Total Duration of Flight	Remarks, Procedures, Maneuvers, Endorsements, PIC Name

I certify that the statements made by me on this form are true.

Pilot's Signature ______________________________

Year 20__ Date	Aircraft Type (Make, Model, Variant)	Aircraft Ident. Mark, Reg. No.	Departure & Arrival		Miles Flown	Flight No.	Number of		No. Instr. Appr.	Aircraft Category and Class								Pilot Function							
			From Time	To Time			TO Day Night	LDG Day Night		SE		ME						PIC (Solo, SPIC, PICUS)		Co-Pilot		Dual Received		Instructor/ Examiner/ Ck. Pilot	
							Page Total																		
							Amount Forward																		
							Total to Date																		

	Operational Conditions					Synthetic Flight				
	Cross-Country	Night	Actual Instrument	Instrument (Hood)		Time of Session	Type of Instruction	Type and Device No.	Total Duration of Flight	Remarks, Procedures, Maneuvers, Endorsements, PIC Name

I certify that the statements made by me on this form are true.

Pilot's Signature ______________________________

Year 20__ Date	Aircraft Type (Make, Model, Variant)	Aircraft Ident. Mark, Reg. No.	Departure & Arrival		Miles Flown	Flight No.	Number of		No. Instr. Appr.	Aircraft Category and Class								Pilot Function							
			From Time	To Time			TO Day Night	LDG Day Night		SE		ME						PIC (Solo, SPIC, PICUS)		Co-Pilot		Dual Received		Instructor/ Examiner/ Ck. Pilot	
						Page Total																			
						Amount Forward																			
						Total to Date																			

	Operational Conditions					Synthetic Flight				
	Cross-Country	Night	Actual Instrument	Instrument (Hood)		Time of Session	Type of Instruction	Type and Device No.	Total Duration of Flight	Remarks, Procedures, Maneuvers, Endorsements, PIC Name

I certify that the statements made by me on this form are true.

Pilot's Signature ______________________________

Year 20__ Date	Aircraft Type (Make, Model, Variant)	Aircraft Ident. Mark, Reg. No.	Departure & Arrival		Miles Flown	Flight No.	Number of		No. Instr. Appr.	Aircraft Category and Class								Pilot Function							
			From Time	To Time			TO Day Night	LDG Day Night		SE		ME						PIC (Solo, SPIC, PICUS)		Co-Pilot		Dual Received		Instructor/ Examiner/ Ck. Pilot	
						Page Total																			
						Amount Forward																			
						Total to Date																			

	Operational Conditions					Synthetic Flight				
	Cross-Country	Night	Actual Instrument	Instrument (Hood)		Time of Session	Type of Instruction	Type and Device No.	Total Duration of Flight	Remarks, Procedures, Maneuvers, Endorsements, PIC Name

I certify that the statements made by me on this form are true.

Pilot's Signature ______________________________

Year 20__ Date	Aircraft Type (Make, Model, Variant)	Aircraft Ident. Mark, Reg. No.	Departure & Arrival From Time	Departure & Arrival To Time	Miles Flown	Flight No.	Number of TO Day Night	Number of LDG Day Night	No. Instr. Appr.	Aircraft Category and Class SE		ME						Pilot Function PIC (Solo, SPIC, PICUS)		Co-Pilot		Dual Received		Instructor/ Examiner/ Ck. Pilot	
						Page Total																			
						Amount Forward																			
						Total to Date																			

		Operational Conditions										Synthetic Flight			Total Duration of Flight		Remarks, Procedures, Maneuvers, Endorsements, PIC Name
		Cross-Country		Night		Actual Instrument		Instrument (Hood)				Time of Session	Type of Instruction	Type and Device No.			

I certify that the statements made by me on this form are true.

Pilot's Signature ______________________________

Year 20__ Date	Aircraft Type (Make, Model, Variant)	Aircraft Ident. Mark, Reg. No.	Departure & Arrival		Miles Flown	Flight No.	Number of		No. Instr. Appr.	Aircraft Category and Class				Pilot Function			
			From / Time	To / Time			TO / Day / Night	LDG / Day / Night		SE	ME			PIC (Solo, SPIC, PICUS)	Co-Pilot	Dual Received	Instructor/ Examiner/ Ck. Pilot
						Page Total											
						Amount Forward											
						Total to Date											

		Operational Conditions										Synthetic Flight					
		Cross-Country		Night		Actual Instrument		Instrument (Hood)				Time of Session	Type of Instruction	Type and Device No.	Total Duration of Flight		Remarks, Procedures, Maneuvers, Endorsements, PIC Name

I certify that the statements made by me on this form are true.

Pilot's Signature ____________________

Year 20__ Date	Aircraft Type (Make, Model, Variant)	Aircraft Ident. Mark, Reg. No.	**Departure & Arrival**		Miles Flown	Flight No.	Number of		No. Instr. Appr.	**Aircraft Category and Class**								**Pilot Function**							
			From Time	To Time			TO Day Night	LDG Day Night		SE		ME						PIC (Solo, SPIC, PICUS)		Co-Pilot		Dual Received		Instructor/ Examiner/ Ck. Pilot	
						Page Total																			
						Amount Forward																			
						Total to Date																			

		Operational Conditions										Synthetic Flight					
		Cross-Country		Night		Actual Instrument		Instrument (Hood)				Time of Session	Type of Instruction	Type and Device No.	Total Duration of Flight		Remarks, Procedures, Maneuvers, Endorsements, PIC Name

I certify that the statements made by me on this form are true.

Pilot's Signature ______________________

Year 20__ Date	Aircraft Type (Make, Model, Variant)	Aircraft Ident. Mark, Reg. No.	Departure & Arrival		Miles Flown	Flight No.	Number of		No. Instr. Appr.	Aircraft Category and Class								Pilot Function							
			From Time	To Time			TO Day Night	LDG Day Night		SE		ME						PIC (Solo, SPIC, PICUS)		Co-Pilot		Dual Received		Instructor/ Examiner/ Ck. Pilot	
						Page Total																			
						Amount Forward																			
						Total to Date																			

	Operational Conditions					Synthetic Flight				
	Cross-Country	Night	Actual Instrument	Instrument (Hood)		Time of Session	Type of Instruction	Type and Device No.	Total Duration of Flight	Remarks, Procedures, Maneuvers, Endorsements, PIC Name

I certify that the statements made by me on this form are true.

Pilot's Signature ______________________________

Year 20__ Date	Aircraft Type (Make, Model, Variant)	Aircraft Ident. Mark, Reg. No.	**Departure & Arrival**		Miles Flown	Flight No.	Number of		No. Instr. Appr.	**Aircraft Category and Class**								**Pilot Function**							
			From Time	To Time			TO Day Night	LDG Day Night		SE		ME						PIC (Solo, SPIC, PICUS)		Co-Pilot		Dual Received		Instructor/ Examiner/ Ck. Pilot	
						Page Total																			
						Amount Forward																			
						Total to Date																			

		Operational Conditions										Synthetic Flight					
		Cross-Country		Night		Actual Instrument		Instrument (Hood)				Time of Session	Type of Instruction	Type and Device No.	Total Duration of Flight		Remarks, Procedures, Maneuvers, Endorsements, PIC Name

I certify that the statements made by me on this form are true.

Pilot's Signature ______________________

Year 20__ Date	Aircraft Type (Make, Model, Variant)	Aircraft Ident. Mark, Reg. No.	**Departure & Arrival**		Miles Flown	Flight No.	Number of		No. Instr. Appr.	**Aircraft Category and Class**								**Pilot Function**							
			From Time	To Time			TO Day Night	LDG Day Night		SE		ME						PIC (Solo, SPIC, PICUS)		Co-Pilot		Dual Received		Instructor/ Examiner/ Ck. Pilot	
						Page Total																			
						Amount Forward																			
						Total to Date																			

		Operational Conditions										Synthetic Flight					
		Cross-Country		Night		Actual Instrument		Instrument (Hood)				Time of Session	Type of Instruction	Type and Device No.	Total Duration of Flight		Remarks, Procedures, Maneuvers, Endorsements, PIC Name

I certify that the statements made by me on this form are true.

Pilot's Signature ______________________________

Year 20__ Date	Aircraft Type (Make, Model, Variant)	Aircraft Ident. Mark, Reg. No.	Departure & Arrival: From / Time	Departure & Arrival: To / Time	Miles Flown	Flight No.	Number of: TO (Day / Night)	Number of: LDG (Day / Night)	No. Instr. Appr.	Aircraft Category and Class: SE	Aircraft Category and Class: ME	Aircraft Category and Class	Aircraft Category and Class	Pilot Function: PIC (Solo, SPIC, PICUS)	Pilot Function: Co-Pilot	Pilot Function: Dual Received	Pilot Function: Instructor/ Examiner/ Ck. Pilot
						Page Total											
						Amount Forward											
						Total to Date											

	Operational Conditions					Synthetic Flight				
	Cross-Country	Night	Actual Instrument	Instrument (Hood)		Time of Session	Type of Instruction	Type and Device No.	Total Duration of Flight	Remarks, Procedures, Maneuvers, Endorsements, PIC Name

I certify that the statements made by me on this form are true.

Pilot's Signature ______________________________

Year 20__ Date	Aircraft Type (Make, Model, Variant)	Aircraft Ident. Mark, Reg. No.	Departure & Arrival		Miles Flown	Flight No.	Number of		No. Instr. Appr.	Aircraft Category and Class								Pilot Function							
			From Time	To Time			TO Day Night	LDG Day Night		SE		ME						PIC (Solo, SPIC, PICUS)		Co-Pilot		Dual Received		Instructor/ Examiner/ Ck. Pilot	
						Page Total																			
						Amount Forward																			
						Total to Date																			

	Operational Conditions					Synthetic Flight				
	Cross-Country	Night	Actual Instrument	Instrument (Hood)		Time of Session	Type of Instruction	Type and Device No.	Total Duration of Flight	Remarks, Procedures, Maneuvers, Endorsements, PIC Name

I certify that the statements made by me on this form are true.

Pilot's Signature ______________________________

Year 20__ Date	Aircraft Type (Make, Model, Variant)	Aircraft Ident. Mark, Reg. No.	Departure & Arrival		Miles Flown	Flight No.	Number of		No. Instr. Appr.	Aircraft Category and Class								Pilot Function							
			From Time	To Time			TO Day Night	LDG Day Night		SE		ME						PIC (Solo, SPIC, PICUS)		Co-Pilot		Dual Received		Instructor/ Examiner/ Ck. Pilot	
						Page Total																			
						Amount Forward																			
						Total to Date																			

	Operational Conditions					Synthetic Flight				
	Cross-Country	Night	Actual Instrument	Instrument (Hood)		Time of Session	Type of Instruction	Type and Device No.	Total Duration of Flight	Remarks, Procedures, Maneuvers, Endorsements, PIC Name

I certify that the statements made by me on this form are true.

Pilot's Signature ______________________________

Year 20__ Date	Aircraft Type (Make, Model, Variant)	Aircraft Ident. Mark, Reg. No.	Departure & Arrival		Miles Flown	Flight No.	Number of		No. Instr. Appr.	Aircraft Category and Class								Pilot Function							
			From Time	To Time			TO Day Night	LDG Day Night		SE		ME						PIC (Solo, SPIC, PICUS)		Co-Pilot		Dual Received		Instructor/ Examiner/ Ck. Pilot	
						Page Total																			
						Amount Forward																			
						Total to Date																			

	Operational Conditions					Synthetic Flight				
	Cross-Country	Night	Actual Instrument	Instrument (Hood)		Time of Session	Type of Instruction	Type and Device No.	Total Duration of Flight	Remarks, Procedures, Maneuvers, Endorsements, PIC Name

I certify that the statements made by me on this form are true.

Pilot's Signature ____________________

Year 20__ Date	Aircraft Type (Make, Model, Variant)	Aircraft Ident. Mark, Reg. No.	Departure & Arrival: From Time	To Time	Miles Flown	Flight No.	Number of: TO Day / Night	LDG Day / Night	No. Instr. Appr.	Aircraft Category and Class: SE		ME						Pilot Function: PIC (Solo, SPIC, PICUS)		Co-Pilot		Dual Received		Instructor/ Examiner/ Ck. Pilot	
						Page Total																			
						Amount Forward																			
						Total to Date																			

	Operational Conditions					Synthetic Flight				
	Cross-Country	Night	Actual Instrument	Instrument (Hood)		Time of Session	Type of Instruction	Type and Device No.	Total Duration of Flight	Remarks, Procedures, Maneuvers, Endorsements, PIC Name

I certify that the statements made by me on this form are true.

Pilot's Signature ______________________

Year 20__ Date	Aircraft Type (Make, Model, Variant)	Aircraft Ident. Mark, Reg. No.	Departure & Arrival		Miles Flown	Flight No.	Number of		No. Instr. Appr.	Aircraft Category and Class								Pilot Function							
			From Time	To Time			TO Day Night	LDG Day Night		SE		ME						PIC (Solo, SPIC, PICUS)		Co-Pilot		Dual Received		Instructor/ Examiner/ Ck. Pilot	
						Page Total																			
						Amount Forward																			
						Total to Date																			

		Operational Conditions										Synthetic Flight			Total Duration of Flight		Remarks, Procedures, Maneuvers, Endorsements, PIC Name
		Cross-Country		Night		Actual Instrument		Instrument (Hood)				Time of Session	Type of Instruction	Type and Device No.			

I certify that the statements made by me on this form are true.

Pilot's Signature ______________________

Year 20__ Date	Aircraft Type (Make, Model, Variant)	Aircraft Ident. Mark, Reg. No.	Departure & Arrival		Miles Flown	Flight No.	Number of		No. Instr. Appr.	Aircraft Category and Class								Pilot Function							
			From / Time	To / Time			TO Day / Night	LDG Day / Night		SE		ME						PIC (Solo, SPIC, PICUS)		Co-Pilot		Dual Received		Instructor/ Examiner/ Ck. Pilot	
							Page Total																		
							Amount Forward																		
							Total to Date																		

	Operational Conditions					Synthetic Flight				
	Cross-Country	Night	Actual Instrument	Instrument (Hood)		Time of Session	Type of Instruction	Type and Device No.	Total Duration of Flight	Remarks, Procedures, Maneuvers, Endorsements, PIC Name

I certify that the statements made by me on this form are true.

Pilot's Signature ______________________________

Year 20__			Departure & Arrival				Number of			Aircraft Category and Class								Pilot Function							
Date	Aircraft Type (Make, Model, Variant)	Aircraft Ident. Mark, Reg. No.	From Time	To Time	Miles Flown	Flight No.	TO Day Night	LDG Day Night	No. Instr. Appr.	SE		ME						PIC (Solo, SPIC, PICUS)		Co-Pilot		Dual Received		Instructor/ Examiner/ Ck. Pilot	
						Page Total																			
						Amount Forward																			
						Total to Date																			

		Operational Conditions										Synthetic Flight					
		Cross-Country		Night		Actual Instrument		Instrument (Hood)				Time of Session	Type of Instruction	Type and Device No.	Total Duration of Flight		Remarks, Procedures, Maneuvers, Endorsements, PIC Name

I certify that the statements made by me on this form are true.

Pilot's Signature ______________________________

Year 20__ Date	Aircraft Type (Make, Model, Variant)	Aircraft Ident. Mark, Reg. No.	**Departure & Arrival**		Miles Flown	Flight No.	Number of		No. Instr. Appr.	**Aircraft Category and Class**								**Pilot Function**							
			From Time	To Time			TO Day Night	LDG Day Night		SE		ME						PIC (Solo, SPIC, PICUS)		Co-Pilot		Dual Received		Instructor/ Examiner/ Ck. Pilot	
							Page Total																		
							Amount Forward																		
							Total to Date																		

	Operational Conditions					Synthetic Flight				
	Cross-Country	Night	Actual Instrument	Instrument (Hood)		Time of Session	Type of Instruction	Type and Device No.	Total Duration of Flight	Remarks, Procedures, Maneuvers, Endorsements, PIC Name

I certify that the statements made by me on this form are true.

Pilot's Signature ____________________

Year			Departure & Arrival				Number of			Aircraft Category and Class								Pilot Function							
20__ Date	Aircraft Type (Make, Model, Variant)	Aircraft Ident. Mark, Reg. No.	From Time	To Time	Miles Flown	Flight No.	TO Day Night	LDG Day Night	No. Instr. Appr.	SE		ME						PIC (Solo, SPIC, PICUS)		Co-Pilot		Dual Received		Instructor/ Examiner/ Ck. Pilot	
							Page Total																		
							Amount Forward																		
							Total to Date																		

	Operational Conditions					Synthetic Flight				
	Cross-Country	Night	Actual Instrument	Instrument (Hood)		Time of Session	Type of Instruction	Type and Device No.	Total Duration of Flight	Remarks, Procedures, Maneuvers, Endorsements, PIC Name

I certify that the statements made by me on this form are true.

Pilot's Signature ______________________________

Year 20__ Date	Aircraft Type (Make, Model, Variant)	Aircraft Ident. Mark, Reg. No.	Departure & Arrival		Miles Flown	Flight No.	Number of		No. Instr. Appr.	Aircraft Category and Class				Pilot Function			
			From Time	To Time			TO Day Night	LDG Day Night		SE	ME			PIC (Solo, SPIC, PICUS)	Co-Pilot	Dual Received	Instructor/ Examiner/ Ck. Pilot
						Page Total											
						Amount Forward											
						Total to Date											

		Operational Conditions										Synthetic Flight					
		Cross-Country		Night		Actual Instrument		Instrument (Hood)				Time of Session	Type of Instruction	Type and Device No.	Total Duration of Flight		Remarks, Procedures, Maneuvers, Endorsements, PIC Name

I certify that the statements made by me on this form are true.

Pilot's Signature ______________________________

Year 20__ Date	Aircraft Type (Make, Model, Variant)	Aircraft Ident. Mark, Reg. No.	Departure & Arrival: From Time	To Time	Miles Flown	Flight No.	Number of: TO Day / Night	LDG Day / Night	No. Instr. Appr.	Aircraft Category and Class: SE	ME			Pilot Function: PIC (Solo, SPIC, PICUS)	Co-Pilot	Dual Received	Instructor/ Examiner/ Ck. Pilot
						Page Total											
						Amount Forward											
						Total to Date											

	Operational Conditions					Synthetic Flight				
	Cross-Country	Night	Actual Instrument	Instrument (Hood)		Time of Session	Type of Instruction	Type and Device No.	Total Duration of Flight	Remarks, Procedures, Maneuvers, Endorsements, PIC Name

I certify that the statements made by me on this form are true.

Pilot's Signature ______________________________

Year 20__ Date	Aircraft Type (Make, Model, Variant)	Aircraft Ident. Mark, Reg. No.	Departure & Arrival		Miles Flown	Flight No.	Number of		No. Instr. Appr.	Aircraft Category and Class								Pilot Function							
			From Time	To Time			TO Day Night	LDG Day Night		SE		ME						PIC (Solo, SPIC, PICUS)		Co-Pilot		Dual Received		Instructor/ Examiner/ Ck. Pilot	
						Page Total																			
						Amount Forward																			
						Total to Date																			

	Operational Conditions					Synthetic Flight				
	Cross-Country	Night	Actual Instrument	Instrument (Hood)		Time of Session	Type of Instruction	Type and Device No.	Total Duration of Flight	Remarks, Procedures, Maneuvers, Endorsements, PIC Name

I certify that the statements made by me on this form are true.

Pilot's Signature ______________________

Year 20__ Date	Aircraft Type (Make, Model, Variant)	Aircraft Ident. Mark, Reg. No.	Departure & Arrival		Miles Flown	Flight No.	Number of		No. Instr. Appr.	Aircraft Category and Class								Pilot Function							
			From Time	To Time			TO Day Night	LDG Day Night		SE		ME						PIC (Solo, SPIC, PICUS)		Co-Pilot		Dual Received		Instructor/ Examiner/ Ck. Pilot	
						Page Total																			
						Amount Forward																			
						Total to Date																			

	Operational Conditions					Synthetic Flight				
	Cross-Country	Night	Actual Instrument	Instrument (Hood)		Time of Session	Type of Instruction	Type and Device No.	Total Duration of Flight	Remarks, Procedures, Maneuvers, Endorsements, PIC Name

I certify that the statements made by me on this form are true.

Pilot's Signature ______________________________

Year 20__ Date	Aircraft Type (Make, Model, Variant)	Aircraft Ident. Mark, Reg. No.	**Departure & Arrival**		Miles Flown	Flight No.	Number of		No. Instr. Appr.	**Aircraft Category and Class**				**Pilot Function**			
			From Time	To Time			TO Day Night	LDG Day Night		SE	ME			PIC (Solo, SPIC, PICUS)	Co-Pilot	Dual Received	Instructor/ Examiner/ Ck. Pilot
						Page Total											
						Amount Forward											
						Total to Date											

		Operational Conditions										Synthetic Flight					
		Cross-Country		Night		Actual Instrument		Instrument (Hood)				Time of Session	Type of Instruction	Type and Device No.	Total Duration of Flight		Remarks, Procedures, Maneuvers, Endorsements, PIC Name

I certify that the statements made by me on this form are true.

Pilot's Signature ______________________

Year 20__ Date	Aircraft Type (Make, Model, Variant)	Aircraft Ident. Mark, Reg. No.	Departure & Arrival		Miles Flown	Flight No.	Number of		No. Instr. Appr.	Aircraft Category and Class								Pilot Function							
			From Time	To Time			TO Day Night	LDG Day Night		SE		ME						PIC (Solo, SPIC, PICUS)		Co-Pilot		Dual Received		Instructor/ Examiner/ Ck. Pilot	
						Page Total																			
						Amount Forward																			
						Total to Date																			

	Operational Conditions					Synthetic Flight				
	Cross-Country	Night	Actual Instrument	Instrument (Hood)		Time of Session	Type of Instruction	Type and Device No.	Total Duration of Flight	Remarks, Procedures, Maneuvers, Endorsements, PIC Name

I certify that the statements made by me on this form are true.

Pilot's Signature ______________________________

Year 20__ Date	Aircraft Type (Make, Model, Variant)	Aircraft Ident. Mark, Reg. No.	Departure & Arrival		Miles Flown	Flight No.	Number of		No. Instr. Appr.	Aircraft Category and Class				Pilot Function			
			From Time	To Time			TO Day Night	LDG Day Night		SE	ME			PIC (Solo, SPIC, PICUS)	Co-Pilot	Dual Received	Instructor/ Examiner/ Ck. Pilot
						Page Total											
						Amount Forward											
						Total to Date											

	Operational Conditions					Synthetic Flight				
	Cross-Country	Night	Actual Instrument	Instrument (Hood)		Time of Session	Type of Instruction	Type and Device No.	Total Duration of Flight	Remarks, Procedures, Maneuvers, Endorsements, PIC Name
										I certify that the statements made by me on this form are true.
										Pilot's Signature ____________________

Year 20__ Date	Aircraft Type (Make, Model, Variant)	Aircraft Ident. Mark, Reg. No.	Departure & Arrival		Miles Flown	Flight No.	Number of		No. Instr. Appr.	Aircraft Category and Class				Pilot Function			
			From Time	To Time			TO Day Night	LDG Day Night		SE	ME			PIC (Solo, SPIC, PICUS)	Co-Pilot	Dual Received	Instructor/ Examiner/ Ck. Pilot
						Page Total											
						Amount Forward											
						Total to Date											

		Operational Conditions										Synthetic Flight					
		Cross-Country		Night		Actual Instrument		Instrument (Hood)				Time of Session	Type of Instruction	Type and Device No.	Total Duration of Flight		Remarks, Procedures, Maneuvers, Endorsements, PIC Name

I certify that the statements made by me on this form are true.

Pilot's Signature ____________________

Year 20__ Date	Aircraft Type (Make, Model, Variant)	Aircraft Ident. Mark, Reg. No.	Departure & Arrival		Miles Flown	Flight No.	Number of		No. Instr. Appr.	Aircraft Category and Class								Pilot Function							
			From Time	To Time			TO Day Night	LDG Day Night		SE		ME						PIC (Solo, SPIC, PICUS)		Co-Pilot		Dual Received		Instructor/ Examiner/ Ck. Pilot	
						Page Total																			
						Amount Forward																			
						Total to Date																			

		Operational Conditions										Synthetic Flight					
		Cross-Country		Night		Actual Instrument		Instrument (Hood)				Time of Session	Type of Instruction	Type and Device No.	Total Duration of Flight		Remarks, Procedures, Maneuvers, Endorsements, PIC Name

I certify that the statements made by me on this form are true.

Pilot's Signature ______________________

Year 20__ Date	Aircraft Type (Make, Model, Variant)	Aircraft Ident. Mark, Reg. No.	**Departure & Arrival**		Miles Flown	Flight No.	Number of		No. Instr. Appr.	**Aircraft Category and Class**								**Pilot Function**							
			From Time	To Time			TO Day Night	LDG Day Night		SE		ME						PIC (Solo, SPIC, PICUS)		Co-Pilot		Dual Received		Instructor/ Examiner/ Ck. Pilot	
						Page Total																			
						Amount Forward																			
						Total to Date																			

	Operational Conditions					Synthetic Flight				
	Cross-Country	Night	Actual Instrument	Instrument (Hood)		Time of Session	Type of Instruction	Type and Device No.	Total Duration of Flight	Remarks, Procedures, Maneuvers, Endorsements, PIC Name

I certify that the statements made by me on this form are true.

Pilot's Signature ____________________

Year			Departure & Arrival				Number of			Aircraft Category and Class				Pilot Function			
20__ Date	Aircraft Type (Make, Model, Variant)	Aircraft Ident. Mark, Reg. No.	From Time	To Time	Miles Flown	Flight No.	TO Day Night	LDG Day Night	No. Instr. Appr.	SE	ME			PIC (Solo, SPIC, PICUS)	Co-Pilot	Dual Received	Instructor/ Examiner/ Ck. Pilot
							Page Total										
							Amount Forward										
							Total to Date										

	Operational Conditions					Synthetic Flight				
	Cross-Country	Night	Actual Instrument	Instrument (Hood)		Time of Session	Type of Instruction	Type and Device No.	Total Duration of Flight	Remarks, Procedures, Maneuvers, Endorsements, PIC Name

I certify that the statements made by me on this form are true.

Pilot's Signature ______________________________

Year 20__ Date	Aircraft Type (Make, Model, Variant)	Aircraft Ident. Mark, Reg. No.	Departure & Arrival		Miles Flown	Flight No.	Number of		No. Instr. Appr.	Aircraft Category and Class								Pilot Function							
			From Time	To Time			TO Day Night	LDG Day Night		SE		ME						PIC (Solo, SPIC, PICUS)		Co-Pilot		Dual Received		Instructor/ Examiner/ Ck. Pilot	
						Page Total																			
						Amount Forward																			
						Total to Date																			

	Operational Conditions					Synthetic Flight				
	Cross-Country	Night	Actual Instrument	Instrument (Hood)		Time of Session	Type of Instruction	Type and Device No.	Total Duration of Flight	Remarks, Procedures, Maneuvers, Endorsements, PIC Name

I certify that the statements made by me on this form are true.

Pilot's Signature ______________________

Year 20__			Departure & Arrival				Number of			Aircraft Category and Class				Pilot Function			
Date	Aircraft Type (Make, Model, Variant)	Aircraft Ident. Mark, Reg. No.	From Time	To Time	Miles Flown	Flight No.	TO Day Night	LDG Day Night	No. Instr. Appr.	SE	ME			PIC (Solo, SPIC, PICUS)	Co-Pilot	Dual Received	Instructor/ Examiner/ Ck. Pilot
						Page Total											
						Amount Forward											
						Total to Date											

		Operational Conditions										Synthetic Flight					
		Cross-Country		Night		Actual Instrument		Instrument (Hood)				Time of Session	Type of Instruction	Type and Device No.	Total Duration of Flight		Remarks, Procedures, Maneuvers, Endorsements, PIC Name
																	I certify that the statements made by me on this form are true.
																	Pilot's Signature ______________________

Year 20__ Date	Aircraft Type (Make, Model, Variant)	Aircraft Ident. Mark, Reg. No.	**Departure & Arrival**		Miles Flown	Flight No.	Number of		No. Instr. Appr.	**Aircraft Category and Class**								**Pilot Function**							
			From Time	To Time			TO Day Night	LDG Day Night		SE		ME						PIC (Solo, SPIC, PICUS)		Co-Pilot		Dual Received		Instructor/ Examiner/ Ck. Pilot	
						Page Total																			
						Amount Forward																			
						Total to Date																			

	Operational Conditions					Synthetic Flight				
	Cross-Country	Night	Actual Instrument	Instrument (Hood)		Time of Session	Type of Instruction	Type and Device No.	Total Duration of Flight	Remarks, Procedures, Maneuvers, Endorsements, PIC Name

I certify that the statements made by me on this form are true.

Pilot's Signature ______________________________

Year 20 ______	Classification of Pilot-In-Command or Co-Pilot Time														
	Single Engine	Single Eng. Over 12,500	Twin Engine	Three Engine	Four Engine	Multi Eng. Over 12,500	Pressurized Aircraft	Turbojet	Turboprop	Helicopter	Communter	On-Demand Operations	Air Carrier		
January															
February															
March															
April															
May															
June															
Total Last 6 mo.															
July															
August															
September															
October															
November															
December															
Total Last 12 mo.															
Amount Forward															
Total To Date															

Physical Examination ______________________ Class __________

Place ______________________

Name of M.D. ______________________

Physical Examination ______________________ Class __________

Place ______________________

Name of M.D. ______________________

Reissuance or Exchange of ______________________

______________ Certificate. New Date ______________

Reissuance or Exchange of ______________________

______________ Certificate. New Date ______________

Reissuance or Exchange of ______________________

______________ Certificate. New Date ______________

Classification of Pilot-In-Command or Co-Pilot Time

Landings		Instrument Approaches		Cross-Country	Actual Instrument	Simulated Instrument	Synthetic Training Instr. Appr.		PIC (Solo, SPIC, PICUS)	Co-Pilot	Dual Received	Flight Instructor	Flight Examiner	Check Pilot		Miles Flown
Day	Night															

Flight Check Date ______________ Type ______________ Examiner ______________

Flight Check Date ______________ Type ______________ Examiner ______________

Flight Check Date ______________ Type ______________ Examiner ______________

Flight Check Date ______________ Type ______________ Examiner ______________

Flight Check Date ______________ Type ______________ Examiner ______________

Flight Check Date ______________ Type ______________ Examiner ______________

Year 20 ______	Classification of Pilot-In-Command or Co-Pilot Time															
	Single Engine	Single Eng. Over 12,500	Twin Engine	Three Engine	Four Engine	Multi Eng. Over 12,500	Pressurized Aircraft	Turbojet	Turboprop	Helicopter	Communter	On-Demand Operations	Air Carrier			
January																
February																
March																
April																
May																
June																
Total Last 6 mo.																
July																
August																
September																
October																
November																
December																
Total Last 12 mo.																
Amount Forward																
Total To Date																

Physical Examination ____________________ Class __________

Place ____________________

Name of M.D. ____________________

Physical Examination ____________________ Class __________

Place ____________________

Name of M.D. ____________________

Reissuance or Exchange of ____________________

____________________ Certificate. New Date __________

Reissuance or Exchange of ____________________

____________________ Certificate. New Date __________

Reissuance or Exchange of ____________________

____________________ Certificate. New Date __________

Classification of Pilot-In-Command or Co-Pilot Time

Landings		Instrument Approaches		Cross-Country	Actual Instrument	Simulated Instrument	Synthetic Training Instr. Appr.		PIC (Solo, SPIC, PICUS)	Co-Pilot	Dual Received	Flight Instructor	Flight Examiner	Check Pilot		Miles Flown
Day	Night															

Flight Check Date ________________ Type ________________ Examiner ________________

Flight Check Date ________________ Type ________________ Examiner ________________

Flight Check Date ________________ Type ________________ Examiner ________________

Flight Check Date ________________ Type ________________ Examiner ________________

Flight Check Date ________________ Type ________________ Examiner ________________

Flight Check Date ________________ Type ________________ Examiner ________________

Year 20 ______	Classification of Pilot-In-Command or Co-Pilot Time														
	Single Engine	Single Eng. Over 12,500	Twin Engine	Three Engine	Four Engine	Multi Eng. Over 12,500	Pressurized Aircraft	Turbojet	Turboprop	Helicopter	Communter	On-Demand Operations	Air Carrier		
January															
February															
March															
April															
May															
June															
Total Last 6 mo.															
July															
August															
September															
October															
November															
December															
Total Last 12 mo.															
Amount Forward															
Total To Date															

Physical Examination ______________________ Class ____________

Place ______________________

Name of M.D. ______________________

Physical Examination ______________________ Class ____________

Place ______________________

Name of M.D. ______________________

Reissuance or Exchange of ______________________

______________ Certificate. New Date ______________

Reissuance or Exchange of ______________________

______________ Certificate. New Date ______________

Reissuance or Exchange of ______________________

______________ Certificate. New Date ______________

Classification of Pilot-In-Command or Co-Pilot Time

Landings		Instrument Approaches		Cross-Country	Actual Instrument	Simulated Instrument	Synthetic Training Instr. Appr.		PIC (Solo, SPIC, PICUS)	Co-Pilot	Dual Received	Flight Instructor	Flight Examiner	Check Pilot		Miles Flown
Day	Night															

Flight Check Date ______________________ Type ______________________ Examiner ______________________

Flight Check Date ______________________ Type ______________________ Examiner ______________________

Flight Check Date ______________________ Type ______________________ Examiner ______________________

Flight Check Date ______________________ Type ______________________ Examiner ______________________

Flight Check Date ______________________ Type ______________________ Examiner ______________________

Flight Check Date ______________________ Type ______________________ Examiner ______________________

Year	Classification of Pilot-In-Command or Co-Pilot Time															
20 ______	Single Engine	Single Eng. Over 12,500	Twin Engine	Three Engine	Four Engine	Multi Eng. Over 12,500	Pressurized Aircraft	Turbojet	Turboprop	Helicopter	Communter	On-Demand Operations	Air Carrier			
January																
February																
March																
April																
May																
June																
Total Last 6 mo.																
July																
August																
September																
October																
November																
December																
Total Last 12 mo.																
Amount Forward																
Total To Date																

Physical Examination ______________________ Class ____________

Place ______________________

Name of M.D. ______________________

Physical Examination ______________________ Class ____________

Place ______________________

Name of M.D. ______________________

Reissuance or Exchange of ______________________

______________________ Certificate. New Date ______________________

Reissuance or Exchange of ______________________

______________________ Certificate. New Date ______________________

Reissuance or Exchange of ______________________

______________________ Certificate. New Date ______________________

Classification of Pilot-In-Command or Co-Pilot Time

Landings		Instrument Approaches		Cross-Country	Actual Instrument	Simulated Instrument	Synthetic Training Instr. Appr.		PIC (Solo, SPIC, PICUS)	Co-Pilot	Dual Received	Flight Instructor	Flight Examiner	Check Pilot		Miles Flown
Day	Night															

Flight Check Date ______________________ Type ______________________ Examiner ______________________

Flight Check Date ______________________ Type ______________________ Examiner ______________________

Flight Check Date ______________________ Type ______________________ Examiner ______________________

Flight Check Date ______________________ Type ______________________ Examiner ______________________

Flight Check Date ______________________ Type ______________________ Examiner ______________________

Flight Check Date ______________________ Type ______________________ Examiner ______________________

Year 20 ______	Classification of Pilot-In-Command or Co-Pilot Time														
	Single Engine	Single Eng. Over 12,500	Twin Engine	Three Engine	Four Engine	Multi Eng. Over 12,500	Pressurized Aircraft	Turbojet	Turboprop	Helicopter	Communter	On-Demand Operations	Air Carrier		
January															
February															
March															
April															
May															
June															
Total Last 6 mo.															
July															
August															
September															
October															
November															
December															
Total Last 12 mo.															
Amount Forward															
Total To Date															

Physical Examination ______ Class ______

Place ______

Name of M.D. ______

Physical Examination ______ Class ______

Place ______

Name of M.D. ______

Reissuance or Exchange of ______

______ Certificate. New Date ______

Reissuance or Exchange of ______

______ Certificate. New Date ______

Reissuance or Exchange of ______

______ Certificate. New Date ______

Classification of Pilot-In-Command or Co-Pilot Time

Landings		Instrument Approaches		Cross-Country	Actual Instrument	Simulated Instrument	Synthetic Training Instr. Appr.		PIC (Solo, SPIC, PICUS)	Co-Pilot	Dual Received	Flight Instructor	Flight Examiner	Check Pilot		Miles Flown
Day	Night															

Flight Check Date ______________________ Type ______________________ Examiner ______________________

Flight Check Date ______________________ Type ______________________ Examiner ______________________

Flight Check Date ______________________ Type ______________________ Examiner ______________________

Flight Check Date ______________________ Type ______________________ Examiner ______________________

Flight Check Date ______________________ Type ______________________ Examiner ______________________

Flight Check Date ______________________ Type ______________________ Examiner ______________________

Year	Classification of Pilot-In-Command or Co-Pilot Time															
20 ______	Single Engine	Single Eng. Over 12,500	Twin Engine	Three Engine	Four Engine	Multi Eng. Over 12,500	Pressurized Aircraft	Turbojet	Turboprop	Helicopter	Communter	On-Demand Operations	Air Carrier			
January																
February																
March																
April																
May																
June																
Total Last 6 mo.																
July																
August																
September																
October																
November																
December																
Total Last 12 mo.																
Amount Forward																
Total To Date																

Physical Examination ______________________ Class ____________

Place ______________________

Name of M.D. ______________________

Physical Examination ______________________ Class ____________

Place ______________________

Name of M.D. ______________________

Reissuance or Exchange of ______________________

______________________ Certificate. New Date ______________________

Reissuance or Exchange of ______________________

______________________ Certificate. New Date ______________________

Reissuance or Exchange of ______________________

______________________ Certificate. New Date ______________________

Classification of Pilot-In-Command or Co-Pilot Time

Landings		Instrument Approaches		Cross-Country	Actual Instrument	Simulated Instrument	Synthetic Training Instr. Appr.		PIC (Solo, SPIC, PICUS)	Co-Pilot	Dual Received	Flight Instructor	Flight Examiner	Check Pilot		Miles Flown
Day	Night															

Flight Check Date ______________________ Type ______________________ Examiner ______________________

Flight Check Date ______________________ Type ______________________ Examiner ______________________

Flight Check Date ______________________ Type ______________________ Examiner ______________________

Flight Check Date ______________________ Type ______________________ Examiner ______________________

Flight Check Date ______________________ Type ______________________ Examiner ______________________

Flight Check Date ______________________ Type ______________________ Examiner ______________________

Year 20 ____	Classification of Pilot-In-Command or Co-Pilot Time															
	Single Engine	Single Eng. Over 12,500	Twin Engine	Three Engine	Four Engine	Multi Eng. Over 12,500	Pressurized Aircraft	Turbojet	Turboprop	Helicopter	Communter	On-Demand Operations	Air Carrier			
January																
February																
March																
April																
May																
June																
Total Last 6 mo.																
July																
August																
September																
October																
November																
December																
Total Last 12 mo.																
Amount Forward																
Total To Date																

Physical Examination ____________________ Class ________

Place ____________________

Name of M.D. ____________________

Physical Examination ____________________ Class ________

Place ____________________

Name of M.D. ____________________

Reissuance or Exchange of ____________________

____________________ Certificate. New Date ____________________

Reissuance or Exchange of ____________________

____________________ Certificate. New Date ____________________

Reissuance or Exchange of ____________________

____________________ Certificate. New Date ____________________

Classification of Pilot-In-Command or Co-Pilot Time

Landings		Instrument Approaches		Cross-Country	Actual Instrument	Simulated Instrument	Synthetic Training Instr. Appr.		PIC (Solo, SPIC, PICUS)	Co-Pilot	Dual Received	Flight Instructor	Flight Examiner	Check Pilot		Miles Flown
Day	Night															

Flight Check Date ______________________ Type ______________________ Examiner ______________________

Flight Check Date ______________________ Type ______________________ Examiner ______________________

Flight Check Date ______________________ Type ______________________ Examiner ______________________

Flight Check Date ______________________ Type ______________________ Examiner ______________________

Flight Check Date ______________________ Type ______________________ Examiner ______________________

Flight Check Date ______________________ Type ______________________ Examiner ______________________

Year 20 ______	Classification of Pilot-In-Command or Co-Pilot Time															
	Single Engine	Single Eng. Over 12,500	Twin Engine	Three Engine	Four Engine	Multi Eng. Over 12,500	Pressurized Aircraft	Turbojet	Turboprop	Helicopter	Communter	On-Demand Operations	Air Carrier			
January																
February																
March																
April																
May																
June																
Total Last 6 mo.																
July																
August																
September																
October																
November																
December																
Total Last 12 mo.																
Amount Forward																
Total To Date																

Physical Examination ______________________ Class ____________

Place ______________________

Name of M.D. ______________________

Physical Examination ______________________ Class ____________

Place ______________________

Name of M.D. ______________________

Reissuance or Exchange of ______________________

______________________ Certificate. New Date ______________________

Reissuance or Exchange of ______________________

______________________ Certificate. New Date ______________________

Reissuance or Exchange of ______________________

______________________ Certificate. New Date ______________________

Classification of Pilot-In-Command or Co-Pilot Time

Landings		Instrument Approaches		Cross-Country	Actual Instrument	Simulated Instrument	Synthetic Training Instr. Appr.		PIC (Solo, SPIC, PICUS)	Co-Pilot	Dual Received	Flight Instructor	Flight Examiner	Check Pilot		Miles Flown
Day	Night															

Flight Check Date ________________ Type ________________ Examiner ________________

Flight Check Date ________________ Type ________________ Examiner ________________

Flight Check Date ________________ Type ________________ Examiner ________________

Flight Check Date ________________ Type ________________ Examiner ________________

Flight Check Date ________________ Type ________________ Examiner ________________

Flight Check Date ________________ Type ________________ Examiner ________________

Year 20 ______	Classification of Pilot-In-Command or Co-Pilot Time														
	Single Engine	Single Eng. Over 12,500	Twin Engine	Three Engine	Four Engine	Multi Eng. Over 12,500	Pressurized Aircraft	Turbojet	Turboprop	Helicopter	Communter	On-Demand Operations	Air Carrier		
January															
February															
March															
April															
May															
June															
Total Last 6 mo.															
July															
August															
September															
October															
November															
December															
Total Last 12 mo.															
Amount Forward															
Total To Date															

Physical Examination ______________________ Class __________

Place ______________________

Name of M.D. ______________________

Physical Examination ______________________ Class __________

Place ______________________

Name of M.D. ______________________

Reissuance or Exchange of ______________________

______________________ Certificate. New Date ______________________

Reissuance or Exchange of ______________________

______________________ Certificate. New Date ______________________

Reissuance or Exchange of ______________________

______________________ Certificate. New Date ______________________

Classification of Pilot-In-Command or Co-Pilot Time

Landings		Instrument Approaches		Cross-Country	Actual Instrument	Simulated Instrument	Synthetic Training Instr. Appr.		PIC (Solo, SPIC, PICUS)	Co-Pilot	Dual Received	Flight Instructor	Flight Examiner	Check Pilot		Miles Flown
Day	Night															

Flight Check Date ______________ Type ______________ Examiner ______________

Flight Check Date ______________ Type ______________ Examiner ______________

Flight Check Date ______________ Type ______________ Examiner ______________

Flight Check Date ______________ Type ______________ Examiner ______________

Flight Check Date ______________ Type ______________ Examiner ______________

Flight Check Date ______________ Type ______________ Examiner ______________

Year 20 ______	Classification of Pilot-In-Command or Co-Pilot Time															
	Single Engine	Single Eng. Over 12,500	Twin Engine	Three Engine	Four Engine	Multi Eng. Over 12,500	Pressurized Aircraft	Turbojet	Turboprop	Helicopter	Communter	On-Demand Operations	Air Carrier			
January																
February																
March																
April																
May																
June																
Total Last 6 mo.																
July																
August																
September																
October																
November																
December																
Total Last 12 mo.																
Amount Forward																
Total To Date																

Physical Examination ______________________ Class __________

Place ______________________

Name of M.D. ______________________

Physical Examination ______________________ Class __________

Place ______________________

Name of M.D. ______________________

Reissuance or Exchange of ______________________

______________________ Certificate. New Date ______________________

Reissuance or Exchange of ______________________

______________________ Certificate. New Date ______________________

Reissuance or Exchange of ______________________

______________________ Certificate. New Date ______________________

Classification of Pilot-In-Command or Co-Pilot Time

Landings		Instrument Approaches		Cross-Country	Actual Instrument	Simulated Instrument	Synthetic Training Instr. Appr.		PIC (Solo, SPIC, PICUS)	Co-Pilot	Dual Received	Flight Instructor	Flight Examiner	Check Pilot		Miles Flown
Day	Night															

Flight Check Date ______________ Type ______________ Examiner ______________

Flight Check Date ______________ Type ______________ Examiner ______________

Flight Check Date ______________ Type ______________ Examiner ______________

Flight Check Date ______________ Type ______________ Examiner ______________

Flight Check Date ______________ Type ______________ Examiner ______________

Flight Check Date ______________ Type ______________ Examiner ______________

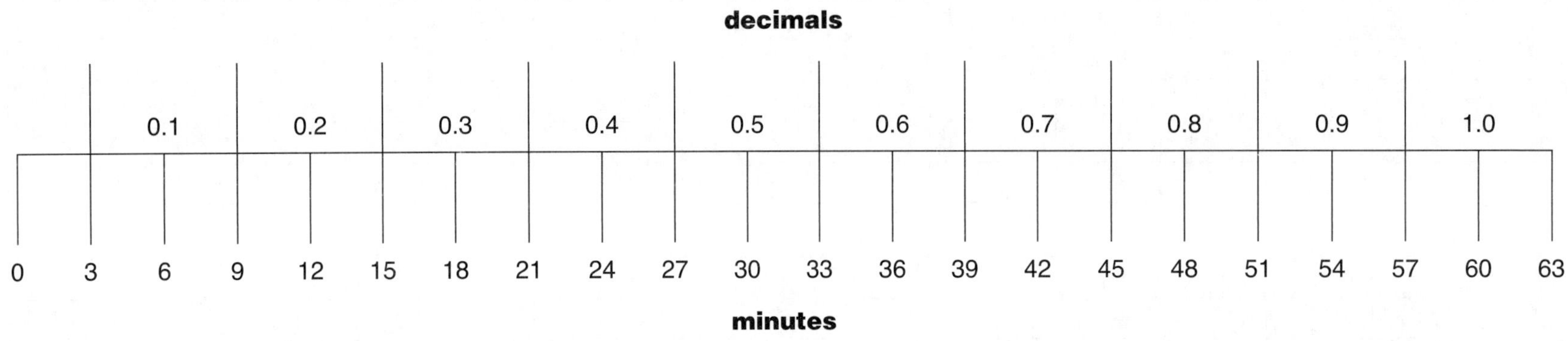
decimals
0.1
0.2
0.3
0.4
0.5
0.6
0.7
0.8
0.9
1.0
0
3
6
9
12
15
18
21
24
27
30
33
36
39
42
45
48
51
54
57
60
63
minutes

INITIAL SOLO ENDORSEMENTS	ADDITIONAL AIRPORT WITHIN 25NM SOLO ENDORSEMENTS
I certify that *(First name, MI, Last name)* ______, has satisfactorily completed the presolo knowledge exam of §61.87(b), received the required presolo training required by §61.87(c), and has demonstrated the proficiency of §61.87(d) and is proficient to make solo flights in *(category, make, model aircraft)* ______. SIGNED ______ DATE ______ CFI NO. ______ EXPIRATION DATE ______	I certify that *(First name, MI, Last name)* ______, has received the required training of § 61.93(b)(1). I have determined that he/she is proficient to practice solo takeoffs and landings at *(airport name)* ______ subject to the following conditions: ______. SIGNED ______ DATE ______ CFI NO. ______ EXPIRATION DATE ______
I certify that *(First name, MI, Last name)* ______, has received the required training to qualify for local solo flying. I have determined he/she meets the applicable requirements of §61.87(n) and is proficient to make solo flights in *(category, make and model aircraft)* ______ until *(maximum 90 days from date given)* ______. *Limitations:* ______. SIGNED ______ DATE ______ CFI NO. ______ EXPIRATION DATE ______	I certify that *(First name, MI, Last name)* ______, has received the required training of § 61.93(b)(1). I have determined that he/she is proficient to practice solo takeoffs and landings at *(airport name)* ______ subject to the following conditions: ______. SIGNED ______ DATE ______ CFI NO. ______ EXPIRATION DATE ______
CROSS-COUNTRY SOLO ENDORSEMENTS	
I certify that *(First name, MI, Last name)* ______, has received the required solo cross-country training and find he/she has met the applicable requirements of §61.93, and is proficient to make solo cross-country flights in a *(category, make and model aircraft)* ______. SIGNED ______ DATE ______ CFI NO. ______ EXPIRATION DATE ______	I have reviewed the cross-country planning of *(First name, MI, Last name)* ______ and find the planning and preparation to be correct to make the solo flight from *(location)* ______ to *(destination)* ______ via *(route of flight)* ______ with landings at *(name the airports)* ______ in a *(make and model aircraft)* ______ on *(date)* ______. *List any applicable conditions or limitations:* ______. SIGNED ______ DATE ______ CFI NO. ______ EXPIRATION DATE ______

ADDITIONAL SOLO ENDORSEMENTS

90-DAY SOLO

I certify that *(First name, MI, Last name)* ________________________________,
has received the required training to qualify for local solo flying. I have determined he/she meets the applicable requirements of §61.87(n) and is proficient to make solo flights in *(category, make and model aircraft)* ____________________
until *(maximum 90 days from date given)* ______________.
Limitations: ________________________________.
SIGNED ________________________ DATE ______________
CFI NO. ________________ EXPIRATION DATE ______________

I certify that *(First name, MI, Last name)* ________________________________,
has received the required training to qualify for local solo flying. I have determined he/she meets the applicable requirements of §61.87(n) and is proficient to make solo flights in *(category, make and model aircraft)* ____________________
until *(maximum 90 days from date given)* ______________.
Limitations: ________________________________.
SIGNED ________________________ DATE ______________
CFI NO. ________________ EXPIRATION DATE ______________

NIGHT SOLO

I certify that *(First name, MI, Last name)* ________________________________,
has received the required presolo training in a *(category, make and model aircraft)*
______________ and determined he/she has demonstrated the proficiency of §61.87(o) and is proficient to make solo flights at night in a *(category, make and model aircraft)* ________________________________.
SIGNED ________________________ DATE ______________
CFI NO. ________________ EXPIRATION DATE ______________

CLASS B SOLO

I certify that *(First name, MI, Last name)* ________________________________,
has received the required training of §61.95(a) and determined he/she is proficient to conduct solo flights in *(name of Class B)* ____________________
airspace. *List any applicable conditions or limitations:* ______________
________________________________.
SIGNED ________________________ DATE ______________
CFI NO. ________________ EXPIRATION DATE ______________

I certify that *(First name, MI, Last name)* ________________________________,
has received the required training of §61.95(a)(1) and determined that he/she is proficient to conduct solo flight operations at *(name of airport)* ______________
____________________. *List any applicable conditions or limitations:*
________________________________.
SIGNED ________________________ DATE ______________
CFI NO. ________________ EXPIRATION DATE ______________

REPEAT X/C SOLO (LESS THAN 50 NM)

I certify that *(First name, MI, Last name)* ________________________________,
has received the required training in both directions between and at both *(airport names)* ____________________ and determined that he/she is proficient in §61.93(b)(2) to conduct repeated solo cross-country flights over that route, which is not more than 50NM from the point of departure, subject to the following conditions: ____________________.
SIGNED ________________________ DATE ______________
CFI NO. ________________ EXPIRATION DATE ______________

INITIAL SPORT PILOT ENDORSEMENTS

SOLO FLIGHT 1st 90-DAY

I certify that *(First name, MI, Last name)* ______________________________,
has received the required training to qualify for local solo flying. I have determined he/she meets the applicable requirements of §61.87(n) and is proficient to make solo flights in *(category, make and model aircraft)* ______________________
until *(maximum 90 days from date given)* ________________.
Limitations: ______________________________________.

SIGNED ______________________________ DATE ______________

CFI NO. ______________________ EXPIRATION DATE ______________

SOLO FLIGHT IN CLASS B, C, AND D AIRSPACE (required each add'l 90-day period)

I certify that *(First name, MI, Last name)* ______________________________,
has received the required training of §61.94(a). I have determined he/she is proficient to conduct solo flights in *(name of Class B, C, or D)* ______________ airspace.
List any applicable conditions or limitations: ______________________.

SIGNED ______________________________ DATE ______________

CFI NO. ______________________ EXPIRATION DATE ______________

TAKING THE AERONAUTICAL KNOWLEDGE TEST

I certify that *(First name, MI, Last name)* ______________________________,
has received the required aeronautical knowledge training of §61.309. I have determined that he/she is prepared for the *(name of knowledge test/aircraft category)*
____.

SIGNED ______________________________ DATE ______________

CFI NO. ______________________ EXPIRATION DATE ______________

TAKING THE SPORT PILOT PRACTICAL TEST

I certify that *(First name, MI, Last name)* ______________________________,
has received the required training of §§61.309 and 61.311 and met the experience requirements of §61.313. I have determined that he/she is prepared for the *(category/class)* ______________________________ practical test.

SIGNED ______________________________ DATE ______________

CFI NO. ______________________ EXPIRATION DATE ______________

I certify that *(First name, MI, Last name)* ______________________________,
has received the required training and demonstrated satisfactory knowledge of deficient areas from the Sport Pilot knowledge test as required per §61.39 for the practical test.

SIGNED ______________________________ DATE ______________

CFI NO. ______________________ EXPIRATION DATE ______________

PASSING THE SPORT PILOT PRACTICAL TEST

I certify that *(First name, MI, Last name)* ______________________________,
has met the training and endorsement requirements of §§61.309, 61.311 and 61.313. I have determined him/her proficient to act as PIC of *(category/class)*
______________________________ of light-sport aircraft.

SIGNED ______________ DATE ______________

PILOT CERT # __________ DPE # __________ EXP. DATE ______________

ADDITIONAL SPORT PILOT ENDORSEMENTS

TAKING FLIGHT PROFICIENCY CHECK FOR AN ADDITIONAL AIRCRAFT

I certify that *(First name, MI, Last name)* ______________________________,
has received the required training of §§61.309 and 61.311. I have determined that he/she is prepared for the *(category / class)* ______________________ ______________________ proficiency check.

SIGNED ______________________ DATE ______________

CFI NO. ______________________ EXPIRATION DATE ______________

PASSING FLIGHT PROFICIENCY CHECK FOR AN ADDITIONAL AIRCRAFT

I certify that *(First name, MI, Last name)* ______________________________,
has met the training and endorsement requirements of §§61.309 and 61.311. I have determined him/her proficient to act as PIC of *(category / class)* ______________________ ______________________ of light-sport aircraft.

SIGNED ______________________ DATE ______________

CFI NO. ______________________ EXPIRATION DATE ______________

PRIVILEGES TO OPERATE A DIFFERENT AIRCRAFT

I certify that *(First name, MI, Last name)* ______________________________,
has received the required training of §61.321 in a *(category / class)* ______________________ ______________________. I have determined him/her proficient to act as PIC of that light-sport aircraft.

SIGNED ______________________ DATE ______________

CFI NO. ______________________ EXPIRATION DATE ______________

AIRSPACE & AIRSPEED ENDORSEMENTS

I certify that *(First name, MI, Last name)* ______________________________,
has received the required training of §61.327 in a *(make and model aircraft)* ______________________. I have determined him/her proficient to act as PIC of a light-sport aircraft that has a V_H greater/less than 87 knots (kts) CAS.

SIGNED ______________________ DATE ______________

CFI NO. ______________________ EXPIRATION DATE ______________

I certify that *(First name, MI, Last name)* ______________________________,
has received the required training of §61.325. I have determined he/she is proficient to conduct operations in Class B, C, or D airspace, at an airport located in Class B, C, or D airspace, or to, from, through, or on an airport having an operational control tower.

SIGNED ______________________ DATE ______________

CFI NO. ______________________ EXPIRATION DATE ______________

I certify that *(First name, MI, Last name)* ______________________________,
has received the training required in accordance with §61.327 in a *(make and model aircraft)* ______________________. I have determined him/her proficient to act as PIC of a light-sport aircraft that has a V_H greater/less than 87 knots (kts) CAS.

SIGNED ______________________ DATE ______________

CFI NO. ______________________ EXPIRATION DATE ______________

RECREATIONAL PILOT ENDORSEMENTS

I certify that *(First name, MI, Last name)* ______________________________,
has received the required training of §61.97(b) and/or I have reviewed the home study curriculum and have determined that he/she is prepared for the *(name of knowledge test/aircraft category)* ______________________________.

SIGNED ______________________________ DATE ______________

CFI NO. ______________________ EXPIRATION DATE ______________

I certify that *(First name, MI, Last name)* ______________________________,
has received the required cross-country training of §61.101(c) and have determined that he/she is proficient in cross-country flying of part 61, subpart E.

SIGNED ______________________________ DATE ______________

CFI NO. ______________________ EXPIRATION DATE ______________

I certify that *(First name, MI, Last name)* ______________________________,
has received the required training of §§61.98(b) and 61.99 and have determined that he/she is prepared for the *(name of practical test)* ______________________________.

SIGNED ______________________________ DATE ______________

CFI NO. ______________________ EXPIRATION DATE ______________

I certify that *(First name, MI, Last name)* ______________________________,
has received the required 180-day recurrent training of §61.101(f) in a *(make and model aircraft)* ______________________. I have determined him/her to be proficient to act as PIC of that aircraft.

SIGNED ______________________________ DATE ______________

CFI NO. ______________________ EXPIRATION DATE ______________

I certify that *(First name, MI, Last name)* ______________________________,
has received the required training of §61.101(b) and have determined he/she is competent to operate at the *(name of airport)* ______________________.

SIGNED ______________________________ DATE ______________

CFI NO. ______________________ EXPIRATION DATE ______________

I certify that *(First name, MI, Last name)* ______________________________,
has received the required training of §61.87 in a *(make and model aircraft)* ______________________. I have determined he/she is prepared to conduct a solo flight on *(date)* ____________ *under the following conditions:* ______________________________.

SIGNED ______________________________ DATE ______________

CFI NO. ______________________ EXPIRATION DATE ______________

PRIVATE PILOT ENDORSEMENTS	COMMERCIAL PILOT ENDORSEMENTS
I certify that *(First name, MI, Last name)* ______________________, has received the required training of §61.105 and/or I have reviewed the home study curriculum and have determined he/she is prepared for the *(name of knowledge test/aircraft category)* ______________________. SIGNED ______________________ DATE ____________ CFI NO. ____________________ EXPIRATION DATE ____________	I certify that *(First name, MI, Last name)* ______________________, has received the required training of §61.125 and/or I have reviewed the home study curriculum and have determined that he/she is prepared for the *(name of knowledge test/aircraft category)* ______________________. SIGNED ______________________ DATE ____________ CFI NO. ____________________ EXPIRATION DATE ____________
I certify that *(First name, MI, Last name)* ______________________, has received the required training of §§61.107 and 61.109 and have determined he/she is prepared for the *(name of practical test)* ______________________ ______________________. SIGNED ______________________ DATE ____________ CFI NO. ____________________ EXPIRATION DATE ____________	I certify that *(First name, MI, Last name)* ______________________, has received the required training of §§61.127 and 61.129 and have determined he/she is prepared for the *(name of practical test)* ______________________ ______________________. SIGNED ______________________ DATE ____________ CFI NO. ____________________ EXPIRATION DATE ____________
I certify that *(First name, MI, Last name)* ______________________, has received the training as required by §61.39(a)(6)(i) within the preceding two calendar months and have determined that he/she is prepared for the *(name of practical test)* ____________________ and has demonstrated satisfactory knowledge of subject areas shown to be deficient on his/her Airman Knowledge Test as required by §61.39(a)(6)(iii). SIGNED ______________________ DATE ____________ CFI NO. ____________________ EXPIRATION DATE ____________	I certify that *(First name, MI, Last name)* ______________________, has received the training as required by §61.39(a)(6)(i) within the preceding two calendar months and have determined that he/she is prepared for the *(name of practical test)* ____________________ and has demonstrated satisfactory knowledge of subject areas shown to be deficient on his/her Airman Knowledge Test as required by §61.39(a)(6)(iii). SIGNED ______________________ DATE ____________ CFI NO. ____________________ EXPIRATION DATE ____________

INSTRUMENT RATING ENDORSEMENTS

I certify that *(First name, MI, Last name)* ______________________________,

has received the required training of §61.65(b) and/or I have reviewed the home study curriculum and have determined that he/she is prepared for the *(name of knowledge test/aircraft category)* ______________________________.

SIGNED ______________________________ DATE ______________

CFI NO. ______________________ EXPIRATION DATE ______________

I certify that *(First name, MI, Last name)* ______________________________,

has received the required training of §61.65(c) and (d) and have determined he/she is prepared for the *(name of practical test)* ______________________________

______________________________.

SIGNED ______________________________ DATE ______________

CFI NO. ______________________ EXPIRATION DATE ______________

I certify that *(First name, MI, Last name)* ______________________________,
has received the training as required by §61.39(a)(6)(i) within the preceding two calendar months and have determined that he/she is prepared for the *(name of practical test)* ____________________________ and has demonstrated satisfactory knowledge of subject areas shown to be deficient on his/her Airman Knowledge Test as required by §61.39(a)(6)(iii).

SIGNED ______________________________ DATE ______________

CFI NO. ______________________ EXPIRATION DATE ______________

INSTRUMENT PROFICIENCY

I certify that *(First name, MI, Last name)* ______________________________,

(pilot certificate) ______________, *(certificate number)* ______________,

has satisfactorily completed the instrument proficiency check required in §61.57(d) in a *(make and model of aircraft)* ______________________________

on *(date)* ______________.

SIGNED ______________________________ DATE ______________

CFI NO. ______________________ EXPIRATION DATE ______________

I certify that *(First name, MI, Last name)* ______________________________,

(pilot certificate) ______________, *(certificate number)* ______________,

has satisfactorily completed the instrument proficiency check required in §61.57(d) in a *(make and model of aircraft)* ______________________________

on *(date)* ______________.

SIGNED ______________________________ DATE ______________

CFI NO. ______________________ EXPIRATION DATE ______________

I certify that *(First name, MI, Last name)* ______________________________,

(pilot certificate) ______________, *(certificate number)* ______________,

has satisfactorily completed the instrument proficiency check required in §61.57(d) in a *(make and model of aircraft)* ______________________________

on *(date)* ______________.

SIGNED ______________________________ DATE ______________

CFI NO. ______________________ EXPIRATION DATE ______________

FLIGHT AND GROUND INSTRUCTOR ENDORSEMENTS

I certify that *(First name, MI, Last name)* ____________________,
has received the required fundamentals of instruction training of §61.185(a)(1).

SIGNED ________________ DATE __________
CFI NO. ________________ EXPIRATION DATE __________

I certify that *(First name, MI, Last name)* ____________________
has received the required training of §61.183(i) and have determined that he/she is competent and proficient in instructional skills for training stall awareness, spin entry, spins, and spin recovery procedures. *(Required of flight instructor applicants for the airplane and glider ratings only.)*

SIGNED ________________ DATE __________
CFI NO. ________________ EXPIRATION DATE __________

I certify that *(First name, MI, Last name)* ____________________,
has received the required training of §61.187(b) and/or I have reviewed the home study curriculum and have determined he/she is prepared for the *(name of knowledge test/aircraft category)* ____________________.

SIGNED ________________ DATE __________
CFI NO. ________________ EXPIRATION DATE __________

I certify that *(First name, MI, Last name)* ____________________
has demonstrated satisfactory proficiency on the appropriate ground instructor knowledge and training subjects of §61.213(a)(3) and (a)(4).

SIGNED ________________ DATE __________
CFI NO. ________________ EXPIRATION DATE __________

I certify that *(First name, MI, Last name)* ____________________,
has received the required CFII training of §61.187(b)(7) and have determined he/she is prepared for the *(name of practical test)* ____________________
____________________.

SIGNED ________________ DATE __________
CFI NO. ________________ EXPIRATION DATE __________

I certify that *(First name, MI, Last name)* ____________________,
has received the training as required by §61.39(a)(6)(i) within the preceding two calendar months and have determined that he/she is prepared for the *(name of practical test)* ____________________ and has demonstrated satisfactory knowledge of subject areas shown to be deficient on his/her Airman Knowledge Test as required by §61.39(a)(6)(iii).

SIGNED ________________ DATE __________
CFI NO. ________________ EXPIRATION DATE __________

ADDITIONAL ENDORSEMENTS

COMPLEX AIRPLANE

I certify that *(First name, MI, Last name)* ______________________________,
(pilot certificate) ________________, *(certificate number)* ________________,
has received the required training of §61.31(e) in a *(make and model of complex airplane)* ______________________________. I have determined that he/she is proficient in the operation and systems of a complex airplane.

SIGNED ______________________________ DATE ______________

CFI NO. ______________________ EXPIRATION DATE ______________

TAILWHEEL

I certify that *(First name, MI, Last name)* ______________________________,
(pilot certificate) ________________, *(certificate number)* ________________,
has received the required training of §61.31(i) in a *(make and model of tailwheel airplane)* ______________________________. I have determined that he/she is proficient in the operation of a tailwheel airplane.

SIGNED ______________________________ DATE ______________

CFI NO. ______________________ EXPIRATION DATE ______________

HIGH-PERFORMANCE AIRPLANE

I certify that *(First name, MI, Last name)* ______________________________,
(pilot certificate) ________________, *(certificate number)* ________________,
has received the required training of §61.31(f) in a *(make and model of high performance airplane)* ______________________________. I have determined that he/she is proficient in the operation and systems of a high performance airplane.

SIGNED ______________________________ DATE ______________

CFI NO. ______________________ EXPIRATION DATE ______________

TYPE RATING

I certify that *(First name, MI, Last name)* ______________________________,
has received the training as required by §61.31(d)(3) to serve as a PIC in a *(category and class of aircraft)* ______________________. I have determined that he/she is prepared to serve as PIC in that *(make and model of aircraft)* ______________________________.

SIGNED ______________________________ DATE ______________

CFI NO. ______________________ EXPIRATION DATE ______________

HIGH ALTITUDE

I certify that *(First name, MI, Last name)* ______________________________,
(pilot certificate) ________________, *(certificate number)* ________________,
has received the required training of §61.31(g) in a *(make and model of pressurized aircraft)* ______________________________. I have determined that he/she is proficient in the operation and systems of a pressurized aircraft.

SIGNED ______________________________ DATE ______________

CFI NO. ______________________ EXPIRATION DATE ______________

ADDED RATING

I certify that *(First name, MI, Last name)* ______________________________,
(pilot certificate) ________________, *(certificate number)* ________________,
has received the required training for an additional *(category and class of aircraft)* ______________________. I have determined that he/she is prepared for the *(name of practical test)* ________________ for the addition of a *(category and class of aircraft)* ______________________________.

SIGNED ______________________________ DATE ______________

CFI NO. ______________________ EXPIRATION DATE ______________

FLIGHT REVIEW

I certify that *(First name, MI, Last name)* ______________________________,

(pilot certificate) ______________, *(certificate number)* ______________,

has satisfactorily completed the flight review required in §61.56(a) on *(date)*

______________.

SIGNED ______________________________ DATE ______________

CFI NO. ______________________ EXPIRATION DATE ______________

I certify that *(First name, MI, Last name)* ______________________________,

(pilot certificate) ______________, *(certificate number)* ______________,

has satisfactorily completed the flight review required in §61.56(a) on *(date)*

______________.

SIGNED ______________________________ DATE ______________

CFI NO. ______________________ EXPIRATION DATE ______________

I certify that *(First name, MI, Last name)* ______________________________,

(pilot certificate) ______________, *(certificate number)* ______________,

has satisfactorily completed the flight review required in §61.56(a) on *(date)*

______________.

SIGNED ______________________________ DATE ______________

CFI NO. ______________________ EXPIRATION DATE ______________

I certify that *(First name, MI, Last name)* ______________________________,

(pilot certificate) ______________, *(certificate number)* ______________,

has satisfactorily completed the flight review required in §61.56(a) on *(date)*

______________.

SIGNED ______________________________ DATE ______________

CFI NO. ______________________ EXPIRATION DATE ______________

I certify that *(First name, MI, Last name)* ______________________________,

(pilot certificate) ______________, *(certificate number)* ______________,

has satisfactorily completed the flight review required in §61.56(a) on *(date)*

______________.

SIGNED ______________________________ DATE ______________

CFI NO. ______________________ EXPIRATION DATE ______________

I certify that *(First name, MI, Last name)* ______________________________,

(pilot certificate) ______________, *(certificate number)* ______________,

has satisfactorily completed the flight review required in §61.56(a) on *(date)*

______________.

SIGNED ______________________________ DATE ______________

CFI NO. ______________________ EXPIRATION DATE ______________